Agent-Based
Modeling and Simulation
with Swarm

T0225423

Chapman & Hall/CRC
Studies in Informatics Series

SERIES EDITOR

G. Q. Zhang

Department of EECS
Case Western Reserve University
Cleveland, Ohio, U.S.A.

PUBLISHED TITLES

Stochastic Relations: Foundations for Markov Transition Systems
Ernst-Erich Doberkat

Conceptual Structures in Practice
Pascal Hitzler and Henrik Schärfe

Context-Aware Computing and Self-Managing Systems
Waltenegus Dargie

Introduction to Mathematics of Satisfiability
Victor W. Marek

Ubiquitous Multimedia Computing
Qing Li and Timothy K. Shih

Mathematical Aspects of Logic Programming Semantics
Pascal Hitzler and Anthony Seda

Agent-Based Modeling and Simulation with Swarm
Hitoshi Iba

Chapman & Hall/CRC
Studies in Informatics Series

Agent-Based Modeling and Simulation with Swarm

Hitoshi Iba

CRC Press
Taylor & Francis Group
Boca Raton London New York

CRC Press is an imprint of the
Taylor & Francis Group, an **informa** business

A CHAPMAN & HALL BOOK

CRC Press
Taylor & Francis Group
6000 Broken Sound Parkway NW, Suite 300
Boca Raton, FL 33487-2742

First issued in paperback 2016

Version Date: 20130429

ISBN 13: 978-1-138-03370-2 (pbk)
ISBN 13: 978-1-4665-6234-9 (hbk)

Visit the Taylor & Francis Web site at
http://www.taylorandfrancis.com

and the CRC Press Web site at
http://www.crcpress.com

Contents

List of Tables

List of Figures

Preface

This book provides a methodology for a multi-agent based modeling approach with the integration of computational techniques, e.g., artificial life, cellular automata, and bio-inspired optimization. The development of such tools contributes to better modeling when addressing tasks from such areas as biology, sociology, civil engineering, economics, granular physics, and art. This attracts attention as a method of understanding complicated phenomena which cannot be solved analytically, in material sciences, ecology, and social science.

In this book, the method of constructing a simulation program is introduced kindly and carefully. Several simulation models with a common theme are treated in each chapter. Each chapter consists of how to assemble the algorithm for realizing a simulation model, a program, a method for visualization, and further research tasks, etc. These models can be automatically learned step by step. While most of the multi-agent simulation is described using the Swarm system, i.e., a commonly used multi-agent simulator, the description is general enough so that the reader can model and develop the same simulation with his own simulator.

The book will provide an overview of multi-agent simulation and supporting materials, organized into seven chapters. Each chapter begins with an overview of the problem and of the current state-of-the-art of the field, and ends with a detailed discussion about multi-agent frameworks. In addition, the appendices provided at the end of this book contain a description of available multi-agent simulation based on such simulators as PSO, ACO, and Swarm systems and source codes available to readers for download.

Chapter 1 provides the background and a basic introduction to simulation and complex systems. It also gives pros and cons of simulation. The limitations of simulation are the subject of lengthy discussions in the field of robotics. We will explain some of these criticisms of simulation.

Chapter 2 provides background and basic introduction to evolutionary computation, i..e., genetic algorithms (GAs), genetic programming (GP), and interactive evolutionary computation (IEC). These are key techniques of simulating multi-agent systems described in the following chapters.

In Chapter 3, we describe the Swarm system, which has been developed by the Santa Fe institute and has been widely used for simulating complex systems. Swarm is a bottom-up model based simulator which consists of biological agents (such as bugs) and abiotic agents (such as a wall or an obstacle) in an artificial world. The motions of the agents are described by simple rules, and the emergence of complex phenomena due to the interaction between the agents is observed. This chapter focuses mainly on the Java library for the Windows edition and gives a tutorial on how to use Swarm, aiming to implement a practical simulation. In later chapters, we will explain the programs of complex systems simulation implemented in Swarm. All software sources described in the text are available from our on-line server for educational purposes.

Chapter 4 describes evolutionary simulation examples by means of GA. Most of them are provided with supplementary demonstration based upon our multi-agent simulation. First, we present a simulation of sexual selection. In this chapter, we present several hypotheses explaining the excessive beauty of certain animals, together with verification experiments based on GAs. This experiment is a remarkable demonstration of the power of female preferences and sexual selection. We also explain an extended version of the prisoner's dilemma known as the "iterated prisoner's dilemma" (IPD), in which the prisoner's dilemma is repeated a number of times with the same participants, and a final score is obtained as the sum of the scores in all iterations. In this section, we first give a detailed discussion of the evolution of cooperation in IPD. Next, we explain the concept of an evolutionarily stable state (ESS), whereby once a certain strategy becomes dominant in a population, invasion by another strategy is difficult. Then we describe an experiment based on research conducted by Cohen, where an IPD strategy evolves using a GA. With these experiments, we have observed that a cooperative relationship emerges in a certain network structure. Subsequently, we give a brief explanation of A-life and its philosophy. Finally, we give a detailed description of remarkable examples, i.e., artificial creatures evolved by Karl Sims.

Marching is a cooperative ant behavior that can be explained by the pheromone trail mode. Many of the ants return to the shorter path, secreting additional pheromones; therefore, the ants that follow also take the shorter path. This model can be applied to the search for the shortest path and is used to solve the traveling salesman problem (TSP) and routing of networks. In Chapter 5, we give a detailed description of ant trail simulation and ant colony optimization (ACO). We also give an advanced topic on the simulation of cooperative behavior of army ants. Army ants build bridges using their bodies along the route from a food source to the nest. Such altruistic behavior helps to optimize the food gathering performance of the ant colony. We present a multi-agent simulation inspired by army ant behavior in which interesting cooperation is observed in the form of philanthropic activity. As a real-world application, we describe an ACO-based approach to solving network routing problems.

In Chapter 6, we describe flocking behaviors of birds or fish (called "boids"). Microscopically, the behavior is very simple and can be modeled using cellular automata; however, macroscopically, the behavior is chaotic and very complex. The models are dominated by interactions between individual birds, and their collective behavior is the result of rules to keep the optimum distance between an individual and its neighbors. Recently researchers designed an effective optimization algorithm using the mechanism behind this collective behavior. This is called particle swarm optimization (PSO), and numerous applications of PSO are reported. In this chapter, details of these methods are provided with several multi-agent simulation examples. Finally, we explain "Swarm chemistry" by Hiroki Sayama. Swarm chemistry is an artificial chemistry framework that employs artificial swarm populations as chemical reactants. Swarm agents steer their motion according to a set of simple kinetic rules, similar to those in boids. We give some simulation results of Swarm chemistry and report on the research results about the application of IEC (interactive evolutionary computation) to evolving desirable motions.

Simulations based on cellular automata (CA) are applied in various fields. These simulations are considered to be an effective method for observing critical behavior in phase transitions. Chapter 7 provides a variety of application examples of cellular automata, e.g., game of life, segregation model, lattice gas automata, Turing model, percolation, traffic simulation, and the Sugarscape model. For instance, we explain the game of life, in which self-organizing capabilities can be used to configure a universal Turing machine. In the area of the game of life, research has been done to find effective rules through GA or GP. The concept of Boolean functions is applied when GP is used. The fitness value is defined by the percentage of correctly processed sequences. Here, the rules obtained using GP were very effective. We give a detailed discussion of this evolutionary learning process. We also discuss the concept of the "edge of chaos" from the behavior of CA proposed by Kauffman and Packard. This concept represents "Class IV" patterns where periodic, aperiodic, and chaotic patterns are repeated. The working hypothesis in artificial life is "life on the edge of chaos." The CA model has been extensively studied for the purpose of simulating emergent properties resulting from multi-agent modeling. We provide a wide range of experimental results based on Swarm in this chapter.

Chapter 8 provides conclusions obtained from the simulation results based on Swarm. We describe a "constructive approach" to "create and understand" complex systems and categorize various multi-agent simulation tests implemented in this book. The significance of bottom-up simulation is analyzed in the summary of the book.

We hope that the multi-agent simulations discussed in this book will contribute to the understanding of complex systems and artificial life.

Hitoshi Iba
Tokyo, Japan

Acknowledgments

To all those wonderful people, the author owes a deep sense of gratitude especially now that this book project has been completed. Especially, the author appreciates the pleasant research atmosphere created by colleagues and students from the research laboratory associated with the Graduate School of

Frontier Sciences and Information Science and Technology at the University of Tokyo.

The author is grateful to his previous group at the Electro-Technical Laboratory (ETL), where he worked for ten years, and to his current colleagues at the Graduate School of Engineering of the University of Tokyo. Particular thanks are due to Dr. Hirochika Inoue and Dr. Taisuke Sato for providing precious comments and advice on numerous occasions. He also wishes to thank Dr. Benedikt Stefansson. The tutorial in Chapter 3 is based on his material from the Swarmfest of 1999. His original version was implemented in Objective-C. We re-implemented this system in Java and modified his materials in accordance with our purpose. However, the author takes full responsibility for any errors in the text and source code.

The author gratefully acknowledges permission from MIT Press to use Figures 4.4, 4.5, 4.6, 4.7, and 4.15, from Prof. Salvacion P. Angtuaco to use Figure 5.12, from Springer-Verlag GmbH to use Figures 5.2 and 5.3, and from Oxford University Press to use Figures 7.1, 7.2, and 7.3.

And last, but not least, he would like to thank his wife Yumiko and his sons and daughter, Kohki, Hirono, and Hiroto, for their patience and assistance.

Chapter 1

Introduction

Suppose that you have to get on an airplane, and that you can choose one of two pilots.

Pilot A: "Although I have 100 hours of flight experience, it is all based on computer simulations. I do not have any real-world experience."

Pilot B: "I have piloted two round trips to the West coast, which means that I have 30 hours of experience."

Which airplane would you get on?

1.1 What is simulation?

A simulation is a modeled experiment. There are other similar-sounding words, such as

Emulation Achievement of an equivalent or better level—for example, the term is used in reference to computer emulators.

Assimilation Incorporation of two or more objects to produce a single entity, or the absorption of objects by other objects—for example, the process of assimilation of carbon dioxide.

"Simulation" is becoming an everyday word. Computer simulations are widely used as conventional procedures for verifying the efficiency of models and processes. This increase in popularity of simulations is attributed to the following advantages.

1. Low cost: simulations are considerably cheaper to implement compared with the cost of setting up an actual experimental facility.

2. Speed: simulated experiments are completed substantially faster than experiments using real-world phenomena.

3. Reproducibility: while real-world verification is difficult to conduct repeatedly, a computer-based experiment can easily be reset and restarted.

1

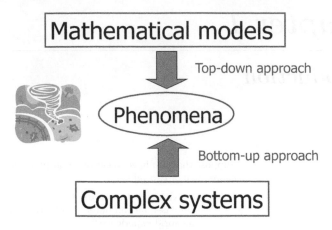

FIGURE 1.1: Bottom-up and top-down approaches.

There are two approaches to the simulation of any phenomenon (Fig. 1.1).

Top-down approach The phenomenon is expressed as a numerical model and simulated by using differential equations or queuing. Such calculations are based mainly on numerical analysis.

Bottom-up approach The phenomenon is expressed as a complex system and simulated through local interaction between elements of the system.

The top-down approach is the conventional simulation procedure where the system under consideration is modeled with equations and the process at each time step is obtained by solving models. However, there is a limit to this process when complex systems are concerned, due to the increasing difficulty in modeling such systems. In reality, the dynamics of the system usually cannot be captured using differential equations. Furthermore, the model equations cannot always be solved analytically and must be solved numerically instead, in which case the effects of numerical errors and noise cannot be neglected.

Therefore, this book discusses bottom-up simulation of complex systems, which is a fundamental paradigm shift in science and engineering. The bottom-up approach for the simulation of complex systems can be applied to a large number of phenomena found, for instance, in science, engineering, the natural world, biological systems, and society. The aim is to model phenomena such as the following:

- Self-organization

- Artificial intelligence

- Evolution

The fundamental principles of a complex system can be summarized as follows:

- The system consists of multiple elements (referred to as agents), forming a multi-agent system.

- The behavior of individual agents is determined by simple rules.

- The rules pertaining to an individual agent describe its behavior in response to local events in its environment, such as encounters with other agents.

- There are no rules that determine the behavior of the system as a whole.

As a result, complex systems display higher-order behavior in comparison to individual-agent systems. Such behaviors are referred to as "emergent properties." A familiar example is the stock market, where individual agents (shareholders) trade shares in accordance with their own rules (heuristics). These rules, such as "buy if the price is expected to increase, sell if the price is expected to decrease," are generated from local information (news, rumors, personal opinions, or memes [59, ch. 2]). Although the stock market is built upon these simple rules, when considered as a whole, the market exhibits emergent fluctuations, for instance, "nervous price trends" or "wait-and-see behavior," leaving the impression that the stock market possesses a will of its own. Emergent properties are explained in Section 4.4.1.

1.2 Simulation of intelligence

There are several intriguing topics of discussion regarding simulation. The most well-known of these is in the field of artificial intelligence (AI).

AI refers to the implementation of intelligence on a computer and is divided into two hypotheses.

Strong AI The viewpoint that true intelligence can be implemented on a computer.

Weak AI The viewpoint that computers can merely give the impression of intelligence.

In the same manner, artificial life (AL), which is mentioned later, can be defined in terms of strong AL and weak AL.

Here, we consider simulation in the sense of "strong AI." More precisely, the rationale behind this approach is that "the appropriately programmed computer really is a mind, in the sense that computers given the right programs can be literally said to understand and have other cognitive states."

However, realizing such a computer is nontrivial, since the definition of "intelligence" is difficult in the first place. Therefore, if a claim is made that

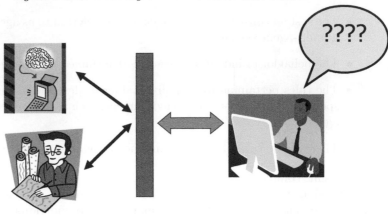

FIGURE 1.2: Turing test.

AI (in the strong sense) has been created, what would be the most appropriate way to evaluate it?

To this end, Alan Turing proposed a test of a machine's capacity to exhibit intelligent behavior, now called the "Turing test," which, despite being powerful, is the subject of numerous disputes (Fig. 1.2). The question of whether machines can think was considered in great depth by Turing, and his final opinion was affirmative. The Turing test can be translated into modern terms in the form of a game involving the exchanging of messages via a discussion board:

- One day, two new users, A and B, join the discussion board.

- When a message is sent to A and B, they both return apt responses.

- Of A and B, one is human and the other is a computer.

- However, it is impossible to determine which is which, regardless of the questions asked.

If a program passes this test (in other words, the computer cannot be identified), the program can be said to simulate intelligence (as long as the questions are valid). A similar contest, named the "Loebner prize" after its patron, the American philanthropist Hugh Loebner, is already held on the Internet.[1] Although an award of US $100,000 and a solid gold medal has been offered since 1990, so far not a single machine participating in the contest has satisfied the criteria for winning.

Nevertheless, a number of problems with the Turing test have been pointed out, and various critical remarks have been issued about potential implementation of AI. A notable example is the challenge to the very "definition of

---

[1]http://www.loebner.net/Prizef/loebner-prize.html

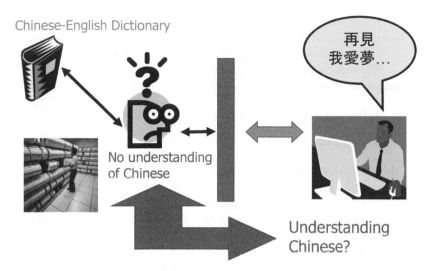

Chinese-English Dictionary

再見
我愛夢...

No understanding
of Chinese

Understanding
Chinese?

FIGURE 1.3: A Chinese room.

intelligence" by John Searle, who upturned the foundations of the Turing test by creating a counter thought experiment. Known as the "Chinese room," Searle's thought experiment can be summarized as follows.

A person is confined to a room with a large amount of written material on the Chinese language (Fig. 1.3). Looking inside the room is impossible, and there are only input and output boxes for submitting sentences and obtaining responses. Having no understanding of Chinese, the person cannot distinguish between different Chinese characters (for this purpose, we assume that the person is British and not Japanese). Furthermore, the person is equipped with a comprehensive manual (written in English) containing rules for connecting sets of Chinese characters. Let us consider that a person who understands Chinese is leading a conversation by inserting questions written in Chinese into the input box and retrieving answers from the output box. Searle provides the following argument.

Suppose that soon the person becomes truly proficient at manipulating Chinese characters in accordance with the instructions, and that the person outside the room also becomes proficient at providing instructions. Then, the answers prepared by the person inside the room would become indistinguishable from answers provided by a Chinese person. Nobody would consider, simply by looking at the provided answers, that the person inside the room does not understand Chinese. However, in contrast to English, in the case of Chinese, the person in the room prepares the answers by formally manipulating characters, without any understanding whatsoever.

It cannot be said that true understanding is achieved simply by looking at the typeface while performing manipulations in accordance with a formal set of rules. However, as demonstrated by the "Chinese room" thought experiment, under specific conditions, human-like behavior can be fabricated by both humans and machines if appropriate formal rules are provided. Searle therefore argues that strong AI is impossible to realize.

Various counterarguments have been considered in response to Searle, and questions that would probably occur to most people include

- Can conversion rules be written for all possible inputs?

- Can such an immense database actually be searched?

However, these counterarguments are devoid of meaning. The former rejects the realization of AI in the first place, and the latter cannot be refuted in light of the possibility that ultra-high-speed parallel computing or quantum computing may exist in the future. Thus, neither one can serve as the basis of an argument.

One powerful counterargument is based on system theory. Although the person in the room certainly lacks understanding, he constitutes no more than a single part of a larger system incorporating other elements, such as the paper and the database, and this system as a whole does possess understanding. This point is integral to the complex systems regarded in this book. The level at which intelligence is sought depends on the observed phenomenon, and if the phenomenon is considered as being an emergent property, the validity of the above system theory can be recognized. Moreover, a debate is ongoing about whether intelligence should be thought of as an integrated concept or as a phenomenon that is co-evolving as a result of evolution.

1.3   Criticism of simulation

It should be kept in mind that a simulation is not an omnipotent tool, which is reflected in the pilots example (see the quote at the beginning of this chapter). The limitations of simulation are the subject of lengthy discussions in the field of robotics.

The ultimate goal in robotics is setting actual machines in motion. However, the process of enabling robots to move is costly, and its implementation is not straightforward. Thus, simulation is actively used for experimental purposes, and an increasing number of studies employ only simulation for conducting experiments, without any verification using actual machines. In conducting research on humanoid robots at our laboratory, an elaborate simulator is always prepared as a preliminary experiment (see Figs. 1.4 and 1.5), and movements realized in the simulator are often impossible to perform with an actual robot. The primary reasons for performing these simulations include locating unforeseen sensor errors, monitoring fatigue due to prolonged

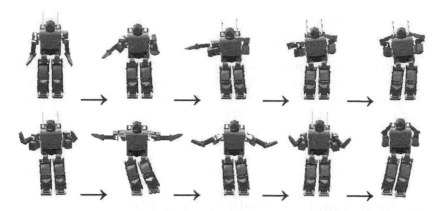

FIGURE 1.4: Simulation of humanoid robot motions.

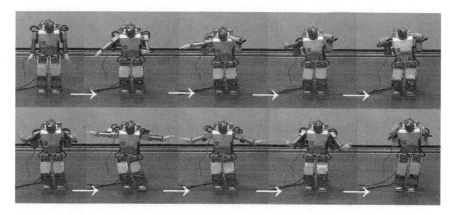

FIGURE 1.5: Real-world motions of a humanoid robot.

use, and studying differences in the friction with the floor. However, the question is whether it would ever be possible to simulate all these circumstances a priori.

In 1991, Rodney Brooks, an eminent scholar in the AI field who proposed concepts such as "intelligence without representation" and "subsumption architecture," presented a highly intriguing argument at the International Joint Conference on Artificial Intelligence (IJCA91). He stipulated that to faithfully recreate the real world in a simulation is impossible, and on the basis of this he emphasized the importance of the following in AI research.

Physical grounding This hypothesis states that intelligence should be grounded in the interaction between a physical agent and its environments.

Embodiment General intelligence cannot be produced in a vacuum. Computers with human intelligence should have a solid sensor-motor base upon which higher cognitive functioning can be built (or evolved).

This statement also serves as a warning regarding both simulation techniques and AI arguments that are likely to become purely theoretical abstractions. With this in mind, simulation experiments must be performed with extreme care.

1.4   Swarm and the Santa Fe Institute

We now present a description of Swarm, which is a multi-agent simulation library developed at the Santa Fe Institute, which is renowned for its research on complex systems. Since Swarm supports Java and Objective-C, it can be easily utilized in object-oriented modeling of phenomena. More specifically, Swarm allows for straightforward implementation of the following functions, each of which is extremely useful in complex systems research.

- Interactive access to fields and methods of objects in the simulation

- Graphical representation of different aspects of the simulation (such as the distribution of agents on a two-dimensional surface and various statistical markers, which can be represented through linear graphs and histograms)

- Assignment of an independent clock (scheduler) to each layer of the simulation

The Santa Fe Institute, which acts as the headquarters for complex systems research, is introduced below. Santa Fe is the capital of the U.S. state of New Mexico, and is located at the center of the state. New Mexico contains a number of settlements known as pueblos that are, even now, populated by Native Americans. Santa Fe is a city that is becoming increasingly popular as a travel destination for American tourists, mainly owing to the preserved landscapes and buildings reminiscent of old times. The famous Mesa Verde National Park is also located nearby, where the escarpments reveal a large number of ruins left behind by a mysterious people, the Anasazi Indians (Fig. 1.6). This tribe thrived as a highly developed civilization about 1400 years ago, and then 700 years later vanished suddenly. This process of thriving and demise is a research topic in Swarm simulations.

The Santa Fe Institute was originally constructed in the ruins of an abbey as part of the Los Alamos National Laboratory. A number of prominent scholars worked at this institute, including Kenneth Arrow (winner of the Nobel Prize in Economics), Murray Gell-Mann (winner of the Nobel Prize in

FIGURE 1.6: Mesa Verde National Park.

Physics for his work on quarks), Robert Axelrod (prisoner's dilemma; see Section 4.3.2), John Holland (the creator of genetic algorithms), and Christopher Langton (who proposed the concept of artificial life).

The "general impossibility theorem" proposed by Arrow is outlined here, since this theorem is related to the concept of complex systems. In this theorem, three bored students (A, B, and C) are discussing which of the following three entertainment options they should choose.

- Movie
- Television
- Karaoke

Let us assume that the order of preference for each person is as follows:

Preference order for A Movie>television>karaoke
.

Preference order for B Television>karaoke>movie
.

Preference order for C Karaoke>movie>television
.

They decide to determine their choice democratically and adopt a majority vote. First, in deciding between a movie and TV, the following distribution determines that a movie is preferable.

```
A movie is preferable to television: A and C.
Television is preferable to a movie: B.
```

∴ Movie (two votes) > television (one vote).

Next, in deciding between television and karaoke, the distribution yields television as the winning choice:

Television is preferable to karaoke: A and B.
Karaoke is preferable to television: C.
∴ Television (two votes) > karaoke (one vote).

From the above results, the order of preference is as follows:

Movie > television > karaoke.

Accordingly, the final decision is to watch a movie. However, looking at the preferences for a movie and karaoke, the distribution leaves karaoke as the winning choice:

A movie is preferable to karaoke: A.
Karaoke is preferable to a movie: B and C.
∴ Karaoke (two votes) > movie (one vote).

In other words, the two results contradict each other.

Arrow generalized this result and demonstrated that serious contradictions arise even in cases with a larger number of people. For example, by considering the relation between the preferences of 100 people, the following result can be obtained by skillfully manipulating the order of preference in a majority vote.

- From a number of majority votes, the overall conclusion is that "x is preferable to y."

- However, only one person prefers x over y.

- The remaining 99 people prefer y over x.

Arrow's general impossibility theorem proved that a "democratic decision" that satisfies the following four criteria does not exist.

1. All preference orders are allowed.

2. Citizen sovereignty is assumed. In other words, x is selected when everyone agrees that x > y.

3. The preference order between two choices depends only on the individual preferences, and is not affected by other alternatives.

4. There is no dictator.

The third criterion is called "independence of irrelevant alternatives," and is an important assumption. The following is a situation where this assumption does not hold.

1. I was ordering lunch at a restaurant.

2. The waitress told me that there was a fish plate and a meat plate, and I ordered the fish plate.

3. The waitress returned a few minutes later and told me that I could also choose pasta.

4. I then ordered the meat plate.

Such a situation actually happened: once, in figure skating where judges score, the existence of a skater who came in fourth place caused the order of the first and second place skaters to be reversed [97].

A more classic example that questions whether democracy is fair is Condorcet's paradox. Condorcet was a mathematician, philosopher, and politician in 18th century France, and is known for investigating the paradox of voting. This paradox appears when voting for three candidates, X, Y, and Z. The result of voting by 60 people was as follows.

- 23 votes for X

- 19 votes for Y

- 18 votes for Z

The question is, should we choose X?
Condorcet clarified that the following paradox exists.
If:

- Z > Y in all 23 people who voted for X

- Z > X in all 19 people who voted for Y

- Y > X in two people, and X > Y in 16 people in a total of 18 people who voted for Z

Then:

- X to Y is 25 to 35, and X to Z is 23 to 37 → X: 0 wins, 2 losses

- Y to X is 35 to 25, and Y to Z is 19 to 41 → Y: 1 win, 1 loss

- Z to X is 37 to 23, and Z to Y is 41 to 19 → Z: 2 wins, 0 losses

Therefore, Z > Y > X, which is the opposite of the vote.
On the other hand, if

- Y > Z in all 23 people who voted for X

- Z > X in 17 people, and X > Z in two people in a total of 19 people who voted for Z

- Y > X in 10 people, and X > Y in eight people in a total of 18 people who voted for Z

Then:

- X to Y is 33 to 27, and X to Z is 25 to 35 → X: 1 win, 1 loss

- Y to X is 27 to 33, and Y to Z is 42 to 18 → Y: 1 win, 1 loss

- Z to X is 35 to 25, and Z to Y is 18 to 42 → Z: 1 win, 1 loss

Therefore, X, Y, and Z become even.

In this way, Arrow proved that the majority vote, which is the fundamental principle in democracy, can be manipulated in an arbitrary manner. For this achievement Arrow received the 4th Nobel Prize in Economics in 1972. The above phenomenon, which appears to undermine the foundations of democracy, can be viewed as the emergence of irrationality in a society consisting of multiple agents. Specifically, although the actions of each individual are rational, this might not necessarily maximize the benefit for society as a whole.

While residing in the United States in 1997, the author had the opportunity to visit the Santa Fe Institute, which left the impression of an extremely open and intellectual environment. During the visit, the author was convinced that innovative research was literally "emerging" from the institute. A notable example of this was the research of Brian Arthur, who proposed the concept of "increased returns," which revolutionized the fundamentals of economics.

Brian Arthur considered economics as a complex system driven by interaction between multiple agents. In conventional economics, common knowledge has dictated that an overall equilibrium exists between supply and demand guided by "the invisible hand," the result of which is that the economy is stable. However, recent trends in financing and stock prices have demonstrated that this assumption is invalid. There are cases where no equilibrium points exist, and the economy as a whole can start to shift in a certain direction as a result of changes in an associated factor. Nothing is capable of stopping this shift once it commences (this process is known as "the bandwagon effect" [59, ch. 2]). A well-known example was competition between videotape standards (Video Home System (VHS) versus Betamax). Although the Betamax format was superior in terms of functionality, VHS eventually prevailed. Thus, if one of two products is considered superior in the spur of the moment, the revenue can start to snowball, even if the actual superiority of the chosen product is unclear.

Hence, a number of complex and unpredictable emergent properties exist in a society constituting an assembly of agents. The mechanisms by which the various phenomena emerge in such complex systems are explained in this book. The next chapter presents an exposition of evolutionary methods, important fundamental techniques (simulation tools) that are used for these explanatory purposes.

Chapter 2

Evolutionary Methods and Evolutionary Computation

> Science is – and how else can I say it – most fun when it plays
> with interesting ideas, examines their implications, and recog-
> nizes that old information might be explained in surprisingly
> new ways. Evolutionary theory is now enjoying this uncom-
> mon vigor (Stephen Jay Gould [44]).

2.1   What is evolutionary computation?

EAs (evolutionary algorithms) and EC (evolutionary computation) are
methods that apply the mechanism of biological evolution to problems in
computer science or in engineering. EAs consider the adaptation process of
organisms to their environments as a learning process. Organisms evolve over a
long period of time by repeating the evolution process whereby species that do
not adapt to the environment become extinct and species that adapt thrive.
This can be applied to practical problems by replacing "environment" with
"problem" or "information that was learned" and "fitness" with "goodness of
the solution."

The most important operators in EAs are selection and genetic operations.
Evolution of individuals does not happen if individuals that adapt better to
their environments are surviving but nothing else is happening. Mutation and
crossover by sexual reproduction result in the generation of a diverse range of
individuals, which in turn promotes evolution. Selection is a procedure where
good individuals are selected from a population. Species that adapt better to
the environment are more likely to survive in nature. The selection procedure
artificially carries out this process.

The typical examples of EAs are genetic algorithms (GAs) and genetic
programming (GP). They are the basic mechanisms for simulating complex
systems. The next sections describe these methods in detail with practical
applications.

2.2   What are genetic algorithms?

GAs have the following characteristics:

- Candidate solutions are represented by sequences of characters

- Mutation and crossover are used to generate solutions of the next generation

Elements that constitute GAs include data representation (genotype or phenotype), selection, crossover, mutation, and alternation of generation. The performance of a search is significantly influenced by how these elements are implemented, as discussed below.

2.2.1   Data representation

The data structure in GAs is either genotype (GTYPE) or phenotype (PTYPE). The GTYPE structure corresponds to chromosomes of organisms (see Fig. 2.1), and is a sequence representing a candidate solution (for example, a bit sequence having a fixed length). This structure is subject to genetic operations such as crossover and mutation. The implementer can design how to convert candidate solutions into sequences. For instance, a GTYPE structure can be obtained by conversion of a candidate solution into a sequence of integers that is then concatenated. On the other hand, PTYPE structures correspond to organisms, and are candidate solutions obtained by interpreting GTYPE structures. The fitness values of candidate solutions are calculated for PTYPE structures.

2.2.2   Selection

The GA is based on the concept of Darwinian evolution, where individuals who adapt better to their environments leave more offspring, while less fit individuals are eliminated. Individuals who adapt to their environments are candidate solutions that are better solutions to a problem, and the measure is the fitness of PTYPE structures.

The following selection methods have been proposed. In particular, the tournament selection is frequently used because scaling is not necessary. In all methods, individuals who have higher fitness are more likely to be selected.

- Roulette selection
  Roulette selection selects individuals with a probability in proportion to their fitness. This is the most general method in EAs; however, procedures such as scaling are necessary to perform searches efficiently.

FIGURE 2.1: GTYPE and PTYPE.

- Tournament selection
  Tournament selection is widely used in EAs. The selection rate in roulette selection is determined by the absolute value of fitness. However, the selection pressure may become too high or too low with roulette selection in problems where the hierarchy of fitness is important but the absolute value is not. Tournament selection uses only the hierarchy of fitness; therefore the above problem does not occur. The computational cost of tournament selection is high because many individuals are selected and fitness values are compared for the number of tournaments.

- Truncate selection
  Individuals are sorted based on fitness and the top $P_s \times M$ individuals (P_s is the selection rate) are selected in truncate selection. The selection pressure is very high; therefore this method is not used in standard GP, but is often used in the estimation of distribution algorithm (EDA), which is an expansion of the GA. The computation cost of this method besides the cost for sorting is very low.

Selection significantly influences the diversity of the population and the speed of convergence; therefore the choice of the selection algorithm and its parameters is very important. For instance, good solutions are often observed with a small number of fitness evaluations by using tournament selection because the selection pressure is very high with a large tournament size. However, the calculations are more likely to quickly converge to and be trapped in an inappropriate local solution.

A different strategy is elitist selection (good individuals are always included in the next generation). The fitness of the best individual never decreases in this strategy with increasing number of generations if the environment against which fitness is measured does not change. However, using elitist selection too early in a search may result in a local solution, or premature convergence.

2.2.3   Genetic operations

When reproduction occurs, the operators shown in Fig. 2.2 are applied to the selected GTYPE to generate a new GTYPE for the subsequent generation. These operators are called GA operators. To keep the explanation simple, we express the GTYPE as a one-dimensional array here. Each operator is analogous to the recombination or mutation of a gene in a biological organism. Generally, the frequency with which these operators are applied, as well as the sites at which they are applied, are determined randomly.

Crossover is an analogy of sexual reproduction where new offspring are generated by combining two parent individuals. There are a number of crossover methods based on the level of granularity in separating each individual, for example, the one-point crossover and the uniform crossover.

The crossover shown in Fig. 2.2 has one crossover point, so it is called a

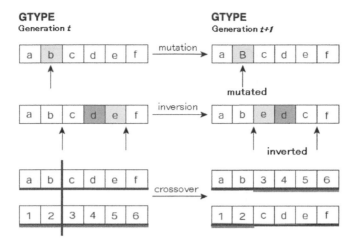

FIGURE 2.2: GA operators.

one-point crossover. Following are some methods for performing the crossover operation.

1. One-point crossover

2. Multi-point crossover (n-point crossover)

3. Uniform crossover

We have already explained the one-point crossover operation (Fig. 2.3(a)). The n-point crossover operation has n crossover points, so if $n = 1$, this is equivalent to the one-point crossover operation. With this crossover method, genes are carried over from one parent alternately between crossover points. A case in which $n = 3$ is shown in Fig. 2.3(b). Two-point crossovers, in which $n = 2$, are often used. Uniform crossovers are a crossover method in which any desired number of crossover points can be identified, so these are realized using a mask for a bit string consisting of $0, 1$. First, let's randomly generate a character string of 0s and 1s for this mask. The crossover is carried out as follows. Suppose the two selected parents are designated as Parent A and Parent B, and the offspring to be created are designated as Child A and Child B. At this point, the genes for offspring Child A are carried over from Parent A when the corresponding mask is 1, and are carried over from Parent B when the mask is 0. Conversely, the genes for offspring Child B are carried over from Parent A when the corresponding mask is 0, and are carried over from Parent B when the mask is 1 (Fig. 2.3(c)).

Mutation in organisms is considered to happen by mutation of nucleotide bases in genes. The GA mimics mutation in organisms by changing the value of a gene location (for example, changing 0 to 1 or 1 to 0). Mutation corresponds

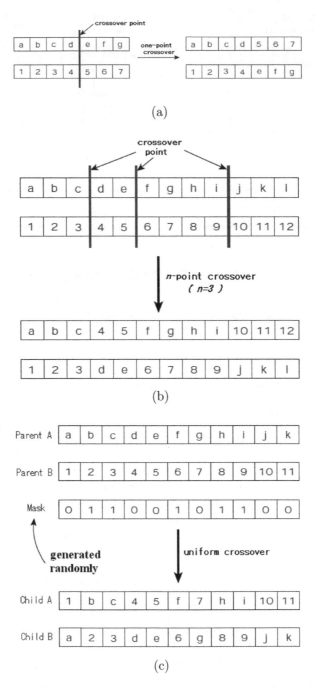

FIGURE 2.3: GA crossovers: (a) one-point crossover, (b) *n*-point crossover, and (c) uniform crossover.

to errors during copying of genes in nature, and is implemented in GAs by changing a character in an individual after crossover (inversion of 0 and 1 in a bit sequence, for instance). Using crossover generally only results in a search of combinations of the existing solutions; therefore using mutation to destroy part of the original gene is expected to increase the diversity of the population and hence to widen the scope of the search. The reciprocal of the length of GTYPE structures is often used as the mutation rate, meaning that one bit per GTYPE structure mutates on average. Increasing the mutation rate results in an increase of diversity, but with the tradeoff of a higher probability of destroying good partial solutions.

2.2.4 Flow of the algorithm

The flow of a GA is as follows:

1. Randomly generate sequences (GTYPE) of the initial population.

2. Convert GTYPE into PTYPE and calculate fitness for all individuals.

3. Select parents according to the selection strategy.

4. Generate individuals in the next generation (offspring) using genetic operators.

5. Check against convergence criteria and go back to 2 if not converged.

Replacing parent individuals with offspring generated through operations such as selection, crossover, and mutation to obtain the population of the next generation is called the alternation of generation. Iterations finish when an individual with an acceptable fitness is found or a predetermined number of generations have been generated. It is also possible to continue the calculations for as long as possible if resources are available, and to terminate when sufficient convergence is achieved or further improvement of fitness appears difficult.

2.2.5 Initialization

All individuals in the first generation are randomly generated individuals in EAs. EAs use genetic operators and therefore have less dependence on the initial state compared to the hill-climbing method, but an extremely biased initial population will decrease the performance of searches. Hence, the initial individuals must be generated to distribute in the search space as uniformly as possible. Figure 2.4 shows an example of a good initialization and a bad initialization. If the triangle △ in Fig. 2.4 is the optimal solution, the optimal solution is expected to be found in a relatively short time from the initial state in Fig. 2.4(a) because the initial individuals are relatively randomly distributed over the search space. On the other hand, the initial distribution in

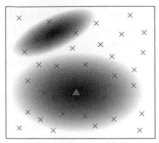 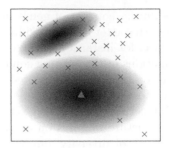

(a) Random initialization   (b) Nonuniform initialization

FIGURE 2.4: Examples of good and bad initialization. The triangles $\triangle$ indicate the optimal solution.

Fig. 2.4(b) has fewer individuals near the optimal solution and more individuals near the local optimal solution; therefore it is more likely that the search will be trapped at the local optimal solution and it will be difficult to reach the optimal solution.

The initial individuals in GAs are determined by randomly assigning 0 or 1 in each gene location, and as the lengths of genes in GAs are fixed, this simple procedure can result in a relatively uniform distribution of initial individuals over the solution space.

2.2.6   Extension of the GA

Many researchers proposed many extensions after John Holland proposed the GA [52], which was the first EA.

The above description considered GTYPE as a sequence with a fixed length; however, modified GAs have been proposed that do not have this restriction. Examples are the real-valued GA that uses GTYPEs of vectors of real numbers and the Messy GA that accommodates sequences of various lengths by pairing the position of a gene and a value. Genetic programming using a tree structure, which is discussed in the next chapter, is another example of a variable length GA. Fitness evaluation methods have also been extended into interactive evolutionary computing (the user determines the fitness and simulates breeding, which allows the GA to be applied in areas where an objective function cannot be explicitly given, such as in designing or the arts; see Section 2.4) and multi-objective optimization (simultaneous optimization of multiple objective functions), and these extensions are known to be very effective in fields like design and the arts.

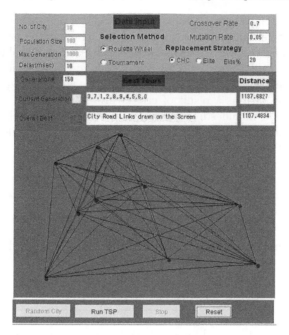

FIGURE 2.5: TSP simulator.

2.2.7   Traveling salesman problem (TSP)

In TSP (traveling salesman problem), there are a number of cities located in different places on a map, and the aim is to look at all of the paths that go through every city exactly once and return to the starting point (called a Hamiltonian cycle or path) and determine the shortest route.

There is no efficient algorithm that will solve the traveling salesman problem; in all cases, an exhaustive investigation is required in order to find the optimum solution. Consequently, as the number (N) of cities grows, we see a dramatic leap in the complexity of the problem. This is called a "combinatorial explosion," and is an important issue (an NP-complete problem) in the field of computer science. The traveling salesman problem is applied in areas such as commodity distribution cost and large scale integration (LSI) pattern technology.

To better understand this problem, let us try using the TSP simulator (Fig. 2.5). Detailed information about this simulator can be found in Appendix A.3.

When using a GA to solve the TSP, the fitness value will be the inverse of the path length, and is defined as follows:

$$Fitness(\text{PTYPE}) = \frac{1}{Length(\text{PTYPE})}, \tag{2.1}$$

where *Length*(PTYPE) is the length of the PTYPE path. As a result, this will be a positive number, and the larger the number, the better.

Now consider how we can solve the traveling salesman problem using a GA. To do this, we will design a GTYPE/PTYPE for this particular problem. If the path is defined as the GTYPE just as it is, we will end up producing points other than the path as a result of crossovers. For example, suppose that we have a route that includes five cities, a, b, c, d, and e. We will assign numbers to these, calling them 1, 2, 3, 4, and 5. Let's pick out the following two paths and examine them.

Name	GTYPE	PTYPE
P_1	13542	$a \to c \to e \to d \to b \to a$
P_2	12354	$a \to b \to c \to e \to d \to a$

Suppose that a crossover occurs between the second and the third cities. This will produce the following:

Name	GTYPE	PTYPE
C_1	12542	$a \to b \to e \to d \to b \to a$
C_2	13354	$a \to c \to c \to e \to d \to a$

This does not solve the traveling salesman problem, i.e., these GTYPEs are not even feasible candidates, because C_1 and C_2 both visit the same city ($2 = b$ and $3 = c$) more than once. This type of GTYPEs (genetic codes) is called "lethal genes." We need to suppress the occurrence of these lethal genes to do an effective search.

The following shows one way to design a GTYPE for a traveling salesman problem. First, we assign a sequence 1,2,3,4,5 to the cities to be visited, which are a, b, c, d, and e. This is a relative sequence in the following sense. Here, the GTYPE of the path called *acedb* (PTYPE) is configured as follows. City "a" is the first city in the above sequence, so we write this as "1." Then we delete "a" from this sequence, leaving *bcde* in the sequence of 1234. In this new sequence, "c" is the second city after "a," so we write "2." We continue in the same way until the GTYPE for *acedb* is found to be 12321.

City	Sequence	Genetic code
a	$abcde \to 12345$	1
c	$bcde \to 1234$	2
e	$bde \to 123$	3
d	$bd \to 12$	2
b	$b \to 1$	1

Using the same method, the GTYPE for the path *abced* will be 11121. Reversing the procedure makes it easy to determine the path from one city to another (PTYPE) from the GTYPE expression. What is important about this GTYPE expression is that the GTYPE obtained as a result of normal

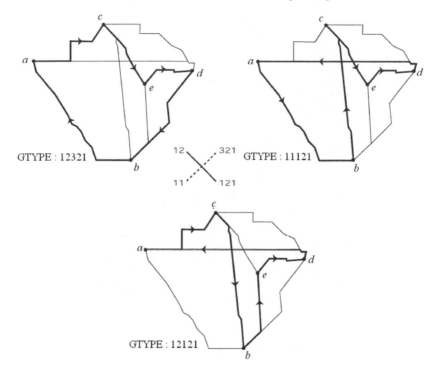

FIGURE 2.6: TSP crossover example.

crossovers indicates the path from one city to the next (Hamiltonian path), i.e., it never becomes a lethal gene. For example, let's consider the crossover we saw earlier (Fig. 2.6).

Name	GTYPE	PTYPE
P_1	12321	$a \to c \to e \to d \to b \to a$
P_2	11121	$a \to b \to c \to e \to d \to a$
C_1	12121	$a \to c \to b \to e \to d \to a$
C_2	11321	$a \to b \to e \to d \to c \to a$

Thus, the GTYPE resulting from the crossover now expresses the path from one city to the next as well. This GTYPE expression is called an "ordinal representation."

Mutations and crossovers are used as GA operators for an ordinal representation. These are basically the same as the operators explained in Section 2.2.3. However, since the GTYPE is expressed as an ordinary character string, rather than a binary expression (a string comprising 0s and 1s), we need to select the gene that will mutate from an appropriate character set.

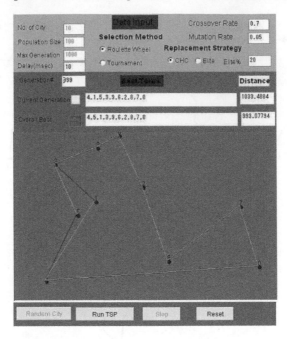

FIGURE 2.7: Results of TSP simulation.

For example, let's consider the GTYPE of the above P_1 (12321). Generally, with ordinal representations, allowed characters for the i-th gene, when the number of cities is N, will be $1, 2, 3, \cdots N - i + 1$. As a result, if the first gene (1) mutates in the above GTYPE, possible characters following the mutation will be 2, 3, 4, and 5. With the second gene (2), they will be 1, 3, and 4.

Figure 2.7 shows the experimental result of a TSP simulator for 10 cities. This TSP simulator can dynamically change the position of cities while searching by the GA. You will observe that the performance decreases temporarily, but the desired solution evolves quickly. This is a characteristic of GAs with population diversity. A population-based search around the current solution allows a somewhat flexible reaction to dynamic changes in environment (change of position of cities). Here, there is no need to search for a new solution from scratch, but improvement of performance from other individuals can be expected. This shows robustness to changes in the environment. See Appendix A.3 for more details.

2.3 What is genetic programming?

Genetic programming (GP, [76]) is one of the most significant extensions for EAs. The key difference between GP and GAs is the description of individuals: the standard GA uses a fixed-length one-dimensional array for an individual, whereas GP uses tree structures. The use of tree structures allows the handling of structural descriptions such as functions and programs that were difficult to handle in GAs. Therefore, GP can be applied to a wide range of applications, and its effectiveness has been confirmed in fields including robotics, circuit design, the arts, and financial engineering. GP has rediscovered inventions that have been accepted as patents, and some results of GP are even superior to technology directly invented by humans [57].

2.3.1 Description of individuals in GP

GP is an expansion of the GA, and one of the main differences is that GP uses tree structures to describe individuals. Each bit has a meaning based on its position in a one-dimensional array with a fixed length that is a description of an individual, which is typically the case in GAs. Therefore partial structures, for instance 01000, do not have any meaning by themselves. Therefore, GAs are applied mainly to problems that search parameters, but there are many limitations in searching structural descriptions such as functions and programs using one-dimensional arrays with a fixed length. GP is an extension designed to search for structural descriptions by using tree structures to describe individuals. GP uses tree structures; therefore partial structures such as $\sin x + z$ have a meaning by themselves. Use of tree structures in GP originates from the use of tree structures as programs (S-expressions) in functional programming, for instance in Lisp. GP can represent S-expressions in Lisp, and therefore GP has strong expressive power because of the ability to represent programs in functional programming languages. GP is known to be Turing complete by adding reading and writing of memory and recursion. General graphs, which are more complex than tree structures, may be used as descriptions, and in fact there is a GP variant that actually uses graphs to describe individuals [114]. However, tree structures usually have sufficient expressive power; therefore standard GP often uses tree structures.

Tree structures are a class of acyclic graphs with nodes and edges. Figure 2.8 is used to explain the components in a tree structure.

- Nodes
  Nodes correspond to the trunk of a tree. The nodes in Fig. 2.9 are $\{A, B, C, D, E\}$.

- Root

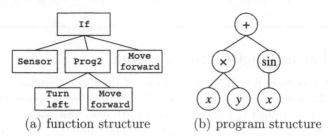

(a) function structure   (b) program structure

FIGURE 2.8: Example of a (a) function and (b) program.

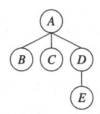

FIGURE 2.9: Example of a tree structure.

This is a node without a parent node. There is one root in each tree structure. The root in Fig. 2.9 is $\{A\}$.

- Terminal nodes
  These are nodes with zero arguments. The terminal nodes in Fig. 2.9 are $T = \{B, C, E\}$.

- Non-terminal nodes
  These are function nodes with one or more arguments. The non-terminal nodes in Fig. 2.9 are $F = \{A, D\}$.

- Child node
  A child node is an argument of another node. The child nodes of node A in Fig. 2.9 are B, C, and D.

- Parent node
  This is the "parent" of a child node: the parent node of node D in Fig. 2.9 is A.

The use of tree structures as descriptions of individuals in GP originates from the use of tree structures as programs in functional programming, as described earlier. Describing programs using one-dimensional arrays, as in GAs, may result in destruction of programs through crossover or mutation. Consider two individuals in GP that are described by

$$x + \sin y$$

and

$$10 \times \sqrt{x - 1}.$$

These can be expressed as linear arrays through the following conversion.

$$+ x \, \sin \, y$$

and

$$\times 10 \, \sqrt{} \, - \, x \, 1.$$

This representation brings the function up front. For example, $x + y$ is expressed as $+ x \, y$.

If the crossover point is chosen between x and sin in the individual above and between $-$ and x in the individual below, the individuals after crossover are

$$+ x \, x \, 1$$

and

$$\times 10 \, \sqrt{} \, - \, \sin \, y.$$

However, these two individuals cannot be evaluated because the syntax is wrong. GP performs genetic operations on partial trees, as discussed later, and therefore can satisfy the requirement that the syntax must be correct.

The use of tree structures allows representation of functions and programs. Fig. 2.8 shows an example of a function (a) and a program (b) that GP can handle. The function in Fig. 2.8(a) represents $x \times y + \sin y$. Functions in general can be represented using a tree structure with appropriate non-terminal nodes and terminal nodes as in the above example. Fig. 2.8(b) is an example of a program for a robot that does the following: the robot has a sensor node that detects obstacles and the robot changes direction if there is an obstacle and moves forward if there is no obstacle. In conclusion, GP can represent a very wide range of knowledge by using tree structures.

As you can see from the above description, the algorithm of GP is almost the same as the GA except that GP uses tree structures as GTYPEs. Thus, we will explain the difference from GA in later sections.

2.3.2 Flow chart of GP

This section describes the typical flow in GP. The following must be decided before using GP when there is a problem to be solved.

- Fitness function

- Nodes to be used

- Design of parameters in the problem

The fitness function evaluates the appropriateness of a solution to the problem. The design of this fitness function can completely change the tendencies in the solutions that will be obtained. Actual examples of the fitness function will be given later.

Which nodes to use, which is the second factor, is important because it determines the size of the solution space. For instance, when the training data in a function regression problem is generated by $\sin x + x$, the solution generated by GP will not be a good approximation of the training data if the non-terminal nodes consist only of $+, -, \times$, and $\div$. On the other hand, using too many nodes would result in the solution space becoming too big. This means that more calculations are necessary before arriving at a solution. Therefore, the nodes to be used must be chosen appropriately, not more and not less.

The third factor is the choice of parameters in GP, and parameters that determine the performance of searches in GP include population size, mutation rate, crossover rate, tournament size (in the case of tournament selection), and maximum tree depth. Searches in a GA are typically carried out with a small number of individuals (~ 100), whereas more individuals are used in GP (generally 1000–10,000 but it depends on the problem). The mutation rate is the ratio of individuals in the population that mutate and is usually about 0.1–0.2. The crossover rate is a similar parameter, and is typically about 0.8. GP uses tree structures and therefore the length is not fixed; however, a limit is usually imposed on the size of the tree structures. The number of nodes increases exponentially in GP due to the bloat phenomenon. Limiting the maximum depth and maximum number of nodes can prevent the generation of excessively large tree structures. Solutions are searched in GP after the above three factors are determined.

Algorithms in GP include the following steps:

1. Random generation of initial population
   M individuals are generated if the number of individuals in the population is M. The initial individuals are generated randomly.

2. Fitness evaluation
   Fitness scores are determined by the fitness function for all M individuals.

3. Selection
   Good solutions are selected through a predetermined selection algorithm.

4. Mutation and crossover
   Mutation and crossover operations are performed on selected good solutions.

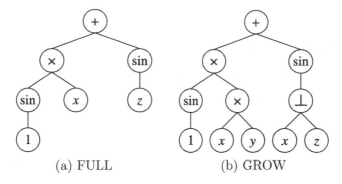

(a) FULL   (b) GROW

FIGURE 2.10: Initialization of a tree structure using FULL and GROW methods.

5. The search ends if the criteria to end the search are satisfied; return to 2 if not.

Each step is described in detail below.

2.3.3   Initialization of tree structures

In contrast to GAs, GP using tree structures, and thus a uniform distribution of initial individuals, is difficult to achieve.

Methods to generate initial individuals in GP usually belong to one of two types, namely "FULL" (full depth) and "GROW" (growing).

- FULL method
  The tree structures can have variable length, but a limit is usually imposed on the maximum depth of tree structures. The FULL method randomly selects from non-terminal nodes until the maximum depth is reached, and then selects from terminal nodes once the maximum depth is reached (Fig. 2.10(a)). Therefore, terminal nodes only exist at the maximum depth in the FULL method.

- GROW method
  In the FULL method, nodes are selected from non-terminal nodes only until the maximum depth is reached; however, in the GROW method, nodes are selected randomly from all nodes until the maximum depth is reached. Once the maximum depth is reached, nodes are randomly chosen from terminal nodes as in the FULL method (Fig. 2.10(b)).

Using the GROW method only or the FULL method only results in biased initial individuals. FULL structures are less likely to be generated when the GROW method is used, and most of the structures that can be generated

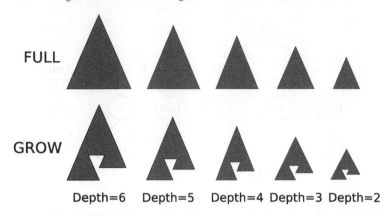

FIGURE 2.11: Initial population obtained with the RAMPED HALF & HALF method.

with the GROW method are not generated when the FULL method is used. Uniformity of tree structures can be defined as follows.

- Uniformity of size
  Groups where most structures have few nodes and groups where most structures have many nodes are not considered to have a uniform distribution over the solution space. Groups where tree structures of various sizes are distributed evenly are more preferable.

- Uniformity of structure
  A population where all individuals are complete trees cannot be considered uniform even though the distribution of size is uniform.

An initialized method called RAMPED HALF & HALF has been proposed, which is a combination of the GROW and FULL methods.

- RAMPED HALF & HALF method
  For a population of M individuals, the population is separated into five groups of $M/5$ individuals each with different depths, e.g., 2, 3, 4, 5, and 6. Half of the individuals in each group are generated with the GROW method, and the other half with the FULL method (Fig. 2.11).

A uniform distribution of initial individuals is important in evolutionary computation because a satisfactory solution cannot be reached if the initial individuals are not uniformly distributed, as discussed before. The RAMPED HALF & HALF method intentionally improves the diversity of the initial individuals, and using this initialization method has been reported to increase the performance of searches.

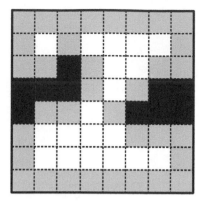

FIGURE 2.12: An example of the wall following problem. The black squares in the figure are walls, and the gray squares are tiles adjacent to walls.

2.3.4 Fitness evaluation

Fitness evaluation is a procedure to quantify how an individual in GP (a tree structure) is adapting to its environment (problem). GP is used to generate programs that determine the motion of robots, and various fitness functions are used depending on the problem in this type of program generation. One famous benchmarking problem in GP is the wall following problem. In this problem, a program to control a robot is searched such that the robot moves as adjacently to a wall as possible in a room with walls, shown in Fig. 2.12. The fitness in this case is

$$fitness_j = \text{(tiles adjacent to a wall that the robot passed)}$$

$$-\text{(tiles away from a wall that the robot passed)}$$

GP finds a program where the robot actually moves along a wall with this fitness evaluation.

The wall following task and its simulator will be described in detail in Section 2.3.6.

2.3.5 Crossover and mutation

Mutation in GP, which uses tree structures, is a natural expansion of mutation in the GA. The most general method is the mutation of partial trees. A node is randomly selected in this mutation method. Next, the partial tree where this node is the root node is replaced with a randomly generated partial tree (Fig. 2.13). The changes from mutation in GAs are relatively small, whereas the changes in GP are large. For instance, the original tree structure becomes a completely new structure if the root node of the original tree was chosen. Therefore, mutation methods with less impact have been proposed,

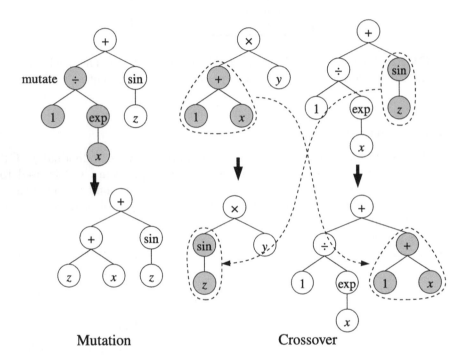

Mutation   Crossover

FIGURE 2.13: Genetic operations in GP. The left is a mutation of a partial tree, and the right is a crossover of partial trees.

one of which is the mutation of nodes. Here, only the selected node is replaced with another node.

The crossover operator in GAs and GP significantly differ from other probabilistic optimization mechanisms. The mutation operator searches nearby structures by slightly changing good solutions. On the other hand, the crossover operator aims to obtain better solutions by combining good solutions. For instance, in a function identification problem to identify x^4, the partial structure x^2 has higher fitness compared to x. In fact, x^2 can be a partial structure of x^4. The concept behind GP is that better solutions can be obtained by repeatedly using crossover to combine good small partial structures called building blocks that formed, for example, through mutation.

Crossover in GAs is an exchange of partial sequences, and crossover in GP is an extension where partial trees are exchanged. Here, partial trees are chosen in two individuals selected through selection, and these selected partial trees are exchanged. Figure 2.13(right) is an example of crossover in GP. $1 + x$ and $\sin z$ are chosen as partial trees in this example. The crossover points are selected at random; however, randomly selecting nodes without distinguishing terminal and non-terminal nodes results in many exchanges of labels between terminal nodes. Terminal nodes cannot be building blocks by themselves; therefore non-terminal nodes are preferentially chosen for crossover in standard GP.

An important condition in genetic operations in GP is that only individuals with correct syntax are always generated when the above crossover or mutation (except mutation of entire nodes) is used. GP uses tree structures to represent genes so that the meaning is not destroyed by genetic operators.

2.3.6 Simulating wall following robots

Let us attempt to evolve a program for controlling a robot using GP. The task here is to obtain a program that causes the robot to follow a wall (this could be useful for controlling a cleaning robot, since dust tends to gather against walls). The robot is capable of four motions: to move forward, move back, turn right, and turn left. It has a group of sensors that provide a 360-degree view. It is placed in an irregular room (i.e., an asymmetric room) and commanded to move about the periphery of the room, following the wall as accurately as possible.

This kind of program is easy to write in Lisp [76]. Let us begin with the adaptive learning necessary to solve this problem with the GP.

The robot is assumed to have 12 sonar sensors (s00–s11) (see Fig. 2.14). The four motions permitted to the robot are straight ahead, straight back, turn left (+30 degrees), and turn right (−30 degrees). Since this GP carries out learning, the terminal nodes T are

$$T = \{\text{S00}, \text{S01}, \cdots, \text{S11}, \Re\}$$

where $\text{S00}, \text{S01}, \cdots, \text{S11}$ are the values output by the 12 distance sensors. $\Re$ is

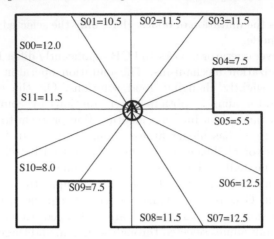

FIGURE 2.14: Robot sensors.

a random number variable (it is generated randomly when initially evaluated, i.e., it is generated once per terminal node). The non-terminal nodes F are

$$F = \{TR, TL, MF, MB, IFLTE, PROGN2\},$$

where TR and TL are functions for turning the robot 30 degrees right and left, respectively, whereas MF and MB are functions for moving the robot 1 foot forward or back, respectively. These functions do not accept arguments. It is assumed that execution of these functions takes 1 time unit, and that the sensor outputs are changed dynamically after execution. Two more functions are incorporated in order for the model to learn the appropriate control relationships. IFLTE (if-less-than-equal) takes four arguments and is interpreted as follows:

(IFLTE a b x y) $\Rightarrow$ if $a \leq b$, then execute x and return the result.
                       if $a > b$, then execute y and the result.

PROGN2 takes 2 arguments, executes them in order and returns the value of the second.

Figure 2.15 presents the wall following simulator by means of GP. The robot moves about the displayed field. Use this simulator to create a program for the desired robot. The robot program will be executed according to a program displaying the GTYPE for a period of 400 time units. The fitness will be determined by how many of the round green circles (tiles) are crossed. Readers should refer to Appendix A.4 for instructions about this simulator.

The generated program always has to cope with even small changes in the environment (such as shifts of the position of the chair in the room). This is what is meant by "robustness" of the program. It is not easy for a human to write this kind of program for a robot; in contrast, GP can perform the necessary searches quite efficiently in order to write such a program.

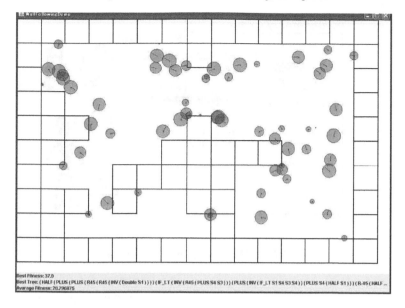

FIGURE 2.15: Wall following by GP.

2.4 What is interactive evolutionary computation?

Let us consider the application of evolutionary computation (EC) to problems such as designing tables that match the atmosphere of a room or composing unobtrusive ringtones for mobile phones.

EC might seem applicable to these problems if they are considered in terms of optimizing the size and color of the boards in the case of the tables, or the frequencies, filters, and other parameters in the case of the synthesizers. However, the problem in this situation is how to evaluate each unit. An evolutionary system based on the survival of the fittest must include a method for evaluating whether individual units are suitable for their environment; in other words, how close they are to the optimal solution. For example, when using GA to evolve to a solution to the shortest path for the traveling salesman problem (TSP), the solution represented by each unit is assigned a degree of fitness in terms of path length (see Section 2.2.7). Is it possible to use a computer in the same manner to determine whether a table matches the atmosphere of a room? Unfortunately, modeling a subjective evaluation process based on human preferences and feelings, and then implementing such a model on a computer, is an extremely difficult task.

For instance, consider the fitness function in the case of a computer drawing a portrait. Conceivably, the first step is to compute the Euclidean distance between the structural information of a face extracted from a photograph and

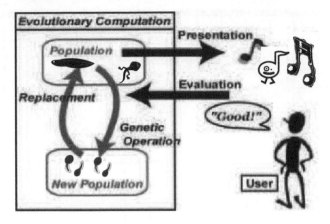

FIGURE 2.16: IEC algorithm.

the portrait drawn by the computer. However, drawing a portrait is not the same as creating a photorealistic representation of a face—rather, drawing a portrait involves the more interesting process of capturing the unique features of the face and then drawing them in a stylized manner. However, can a computer determine whether a person is identical in a facial photograph and a portrait in which specific facial features are captured and stylized? This fitness problem is considered to be extremely difficult to implement using the existing level of technology.

Nevertheless, there is something close to all of us that is capable of performing such evaluations instantaneously: the brain. Humans have evolved a method for direct evaluation of individual units, and in this regard, the human evaluation system can be incorporated into an optimization system as an evaluation function. Thus, when EC is employed for optimization based on human subjective evaluation, this process is then referred to as interactive evolutionary computation (IEC) [10, 116].

Put simply, IEC is EC in which a human is substituted for the fitness function. In IEC, the person using the system (the user) directly evaluates each unit, and the viability (fitness) of subsequent generations is derived depending on user preferences (Fig. 2.16). When implementing this system, personal preferences and sensations are not modeled, and evaluation based on user subjectivity is incorporated into the system as a black box. In contrast to conventional EC, where evolution is modeled as the result of a struggle for survival, IEC is inspired by the process undertaken by humans of intentionally breeding agricultural products and domestic animals.

Modern keywords such as "emotional (kansei) engineering" and "humanized technology" reveal the increasing tendency for technology to make use of human subjectivity. As part of this trend, IEC is beginning to attract attention as a method able to incorporate the user's subjective evaluation system.

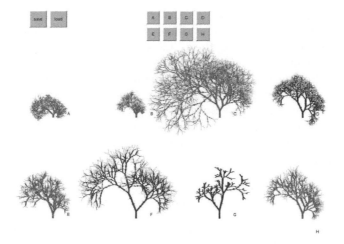

FIGURE 2.17: CG synthesis of plants based on L-systems.

Various types of human evaluation scales are used in IEC. For example, psychological experiments often use a scale with at most 7 grades, although a scale with 10 or 100 grades would certainly increase clarity in everyday situations. In contrast, for two-step evaluation, members of the population can be either selected (or not) as a parent for the next generation. This process is referred to as "simulated breeding" [121], since it is analogous to artificial breeding, in which desirable features are selected and then bred.

In his 1986 book *The Blind Watchmaker* [24], Richard Dawkins, the well-known author of *The Selfish Gene*, describes how images created according to simple rules can evolve into extremely complex and intriguing images as a result of rule mutation and user selection. These image sequences were called *biomorphs*, since they appear to resemble animals (e.g., insects) at first sight. Biomorphs can be regarded as the first example of IEC, specifically, evolution artificially created with a computer through selection based on human subjective preferences.

Hence, IEC offers a new creative technique based on user selections. Since many artists and researchers were attracted to this method after being captivated by biomorphs, the initial focus of IEC research was in artistic fields, particularly the application of IEC in computer graphics (CG) [106]. Such applications cover a number of areas, including CG synthesis of plants based on L-systems (Fig. 2.17, see also Appendix A.5), portrait synthesis, graphic art, three-dimensional (3D) CG synthesis for virtual reality, avatar motion design [122], and animation [10]. Below, we focus on graphic art and design as a representative IEC application.

Simulated Breeding for ART (SBART) [121] uses genetic programming (GP) to create 2D images and is an example of an IEC-based graphic art system. Drawing in SBART is performed on the basis of a set of equations defining 3D vector operations on variables x and y. Beautiful images are cyclically generated by substituting the x- and y-coordinates of the current image's pixels into the equations and transforming the resulting values back into image information. Specifically, each vector component obtained through the substitution process is colored in hue–saturation–brightness space, and movies can be generated from images by introducing time, t, as a separate variable.

A simulator named *LGPC for Art* (see Fig. 2.18) is currently being developed by the author's laboratory, which uses SBART as a reference platform. It has the following basic procedure.

Step 1 Click Clear if you do not like any of the pictures in the 20 windows. This re-initializes all of the windows.

Step 2 Click one of the pictures to select it if you like it. The frame will turn red. You can choose as many as you like.

Step 3 Press the OK button to display the population of the next generation of genes produced from the image you selected.

Step 4 Repeat from **Step** 1 to **Step** 3.

Any image you like can be stored (with the Gene_Save command) and any stored image can be loaded (with the Gene_Load command) to replace the currently displayed image. The 20 original windows loaded with the View tab are numbered (#1 is at the upper left, and numbering proceeds left to right, then up to down).

With this simulator, users can experience the excitement of "breeding" custom images. In fact, the author used this system to design the cover of his own book, and the design process of creating a camera-ready image, which would normally require several days for a graphic designer, was completed within several hours.

Refer to Appendix A.6 for instructions about the installation and usage of LGPC for Art. Play with this process to "cultivate" new images.

The main purpose of the above mentioned systems is the generation of aesthetically pleasing CG. However, another viewpoint is to focus on the human–computer interaction that occurs during the IEC process (between computer generation of units and user selections).

Karl Sims, one of the pioneers of IEC-based graphic art, developed an IEC system that in itself was a work of art. The system was named *Galápagos* and consisted of 12 monitors, each with a foot sensor. Individual monitors displayed computer-simulated virtual creatures with abstract morphologies and independent movements. Visitors were able to choose the parents for the next generation by stepping on the sensors of the monitors displaying their preferred creatures. Through repetition of this process of heterogenesis, the simulator could evolve virtual creatures pleasing to the eye able to perform intriguing movements.

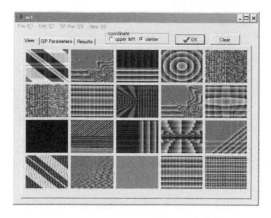

FIGURE 2.18: An overview of LGPC for art.

The developed virtual creatures carried genotypes (symbolic expressions) that were considerably more complex and incomprehensible than Sims had ever envisaged. The interaction between artist, computers, and visitors gave rise to forms that were impossible for either the artist or computers to independently create. One of the difficult aspects of biology is that humans know only about life based on DNA (life as we know it); however, by considering evolutionary mechanisms, this installation showed forms that could potentially exist (life as it could be), and presented an extremely interesting view of artificial life (see also Section 4.4.1).

IEC has also been applied to music through the celebrated MIDI-based jazz improvisation system, *GenJam* [12]. GenJam uses a population of double-layered genotypes in which one layer contains units consisting of eight ordered digitized musical notes and the other layer contains units that determine the arrangement of the notes. These two layers correspond to bars and musical phrases consisting of sequences of bars, respectively. Following a predefined rhythm section and the code progression, GenJam plays a melody by mixing the two sets of units. An evaluator presses "g" (good) if he likes the melody and "b" (bad) otherwise, and so assigns a score to the corresponding units. Selection and heterogenesis are thus iterated on the basis of these scores, and the generated bars and phrases are continuously improved (or, at the least, become closer to the evaluator's preferences). Conceivably, this process corresponds to jazz musicians building a melody through repeated trial and error.

Al Biles, the creator of GenJam and a veteran jazz player, performs jam sessions with GenJam trained in this way. At times, GenJam is capable of creating surprises since it presents renditions that a human would rarely consider. Hence, the system represents another example of this new form of human–computer interaction.

FIGURE 2.19: An example rock rhythm generated by *CONGA*.

Undertaking similar research, the author has developed a rhythm generation system named *CONGA* [120]. A characteristic feature of this system is
that the search technique is based on a combination of a GA and GP. First,
short rhythms with a length of one bar are represented as the character sequences of genes, which are then used by a GA. Conversely, GP uses functions
such as "repeat" to arrange the rhythms represented by GA units, with the
aim of realizing a musical structure. In this way, more pleasant musical pieces
can be generated by combining the short gene expressions implemented using
a GA and the structural representation implemented using GP. Figure 2.19
shows an example rock rhythm generated with CONGA. Sample tunes generated with this system can be downloaded from the author's website.

Other practical examples of IEC application can be given for various design fields, such as bridge construction, the automotive industry, and fashion.
Caldwell et al. [14] applied IEC to the preparation of composite portraits
(montages) by using an interactive GA to optimize parameters, such as the
shapes and positions of the eyes, mouth, and other facial features. Generally,
the recollection of an eyewitness tends to be vague, and they experience difficulty conveying their impression of a face. However, by using Caldwell et al.'s
system, witnesses can generate portraits by repeatedly selecting the face most
closely resembling that of the suspect. In fact, Caldwell et al. created this
system as a result of his frustration after an incident in which 20 U.S. dollars
was stolen from him during a drug store robbery. The portraits of the robbers
that Caldwell et al. constructed with this system were extremely precise, and
consequently they were arrested without difficulty.

Although IEC was initially applied to subjective fields, such as the arts,
the use of IEC has recently been extended to domains where objective criteria are needed. Explicitly, IEC has been applied to "humanized technology,"
in which the aim is to integrate human sensations and subjectivity into systems. We explain IEC-based automatic adjustment of digital hearing aids as
a representative example of such applications [115].

Hearing aids provide assistance by increasing the sound pressure level of
frequencies that are difficult for the user to hear. To maximize the capabilities
of a hearing aid, it must be adjusted to match the user's hearing characteris-

tics. In the conventional adjustment method, hearing test results are compared with the typical audible frequency range for humans, and parameters of the hearing aid are adjusted by professional doctors and technicians. In addition to being rather time-consuming, this method is limited in terms of the hearing characteristics that can be measured. Moreover, in reality, only the user can determine their hearing ability, which cannot be adequately determined even by the most capable doctors and technicians.

Hence, research is being conducted on IEC systems where users adjust the parameter levels of the hearing aid signal based on their own hearing [115]. In this method, a "compensation surface" that represents the input sound pressure, frequency, and compensation level is constructed by combining 3D Gaussian surfaces. The parameters of these Gaussian surfaces are optimized interactively using a GA. According to the results of a user survey, about 70% of users ranked hearing aids optimized by IEC as considerably improved over hearing aids optimized by the conventional method.

In recent years, *kansei* (affective) engineering, which utilizes human emotions, has also enjoyed considerable attention. As in the above mentioned case of adjusting hearing aids, the most anticipated fields for application of IEC are engineering applications that utilize the sensations and preferences of the user, which are difficult to measure quantitatively.

As we have seen above, IEC allows the combining of computational power and the evaluative abilities of humans in order to perform "optimization" and "retrieval" based on user subjectivity and preferences. Nevertheless, IEC suffers from problems that do not arise in standard EC. First, the psychological burden placed on users to evaluate units is rather heavy. IEC users are required to perform the repetitive mechanical operations of comparing and evaluating a large number of units and assigning fitness scores to numerous generations. Such repetitive work is likely to induce psychological discomfort and fatigue. Therefore, an important challenge in IEC research is the reduction of this burden. For this purpose, various human–computer interaction setups (the process of inputting evaluations and displaying units using an interface) have been devised. Devising such setups is an extremely important research topic, considering the expected wide application of IEC (refer to Takagi and Ohsaki; [115] for details). Another problematic aspect is the evaluation of the system and the results obtained. This issue is a challenge not only for IEC but also for all systems in which subjective evaluation is utilized. Solution requires the development of a framework that allows for investigating both the system operability and the resultant content.

In the future, IEC is expected to develop as a fundamental technology, enabling completely new methods of human–computer interaction.

2.4.1   Interactive music composition based on Swarm simulation

Ando and the author implemented an interactive music creation by linking Swarm and external software [8]. This musical product imports the MIDI signals of the piano performance of humans, and reflects it in the state of multi-agent simulation constructed in Swarm. Furthermore, the backward conversion, that is, creating a piano performance from the simulation environment and the conversion of actual piano performance, are also carried out. These conversions are done in real time while the music is being played. A human performer and the swarm intelligence in Swarm influence each other and perform alternatively. Performances like 4 bar exchange jazz can be created. Using the MIDI piano (in which the computer's signals move the keyboard and produce sound) of the Yamaha Company, a real piano performance is created in the computer. This way, a kind of interaction is created as if a human performer and Swarm are doing a duet on a single piano (see Fig. 2.20). The computer for Swarm is connected to the MIDI piano through the computer for music. Both the flows from Swarm to piano and from piano to swarm are realized. In the connection between the computers, to ensure the real-time characteristic, the network protocol called OpenSoundControl is used. OpenSoundControl is extended such that it can be given from Swarm as well. Apart from interactive performances using the game of life, sound effects were also created based on the state transition of boid and pheromone trail. Musical performances using the Swarm system are actually being done at concerts and have been receiving positive feedback.

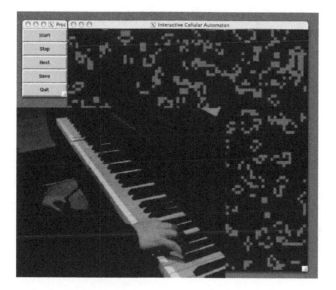

FIGURE 2.20: A four-handed performance on the piano and Swarm.

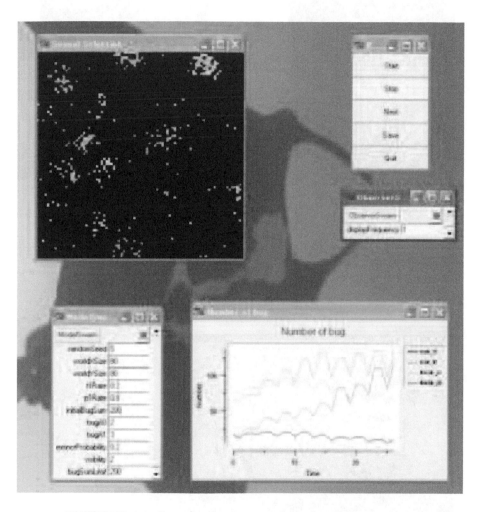

FIGURE 4.9: Sexual selection in a two-dimensional space.

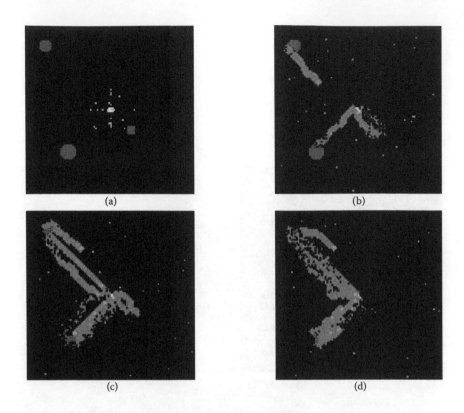

(a)   (b)

(c)   (d)

FIGURE 5.4: Pheromone trails of ants.

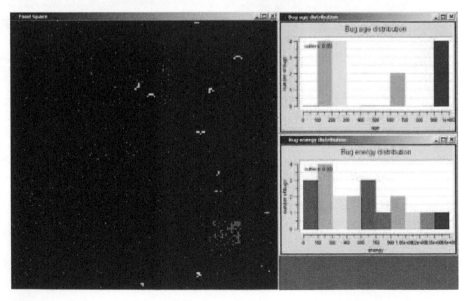

FIGURE 6.32: BUGS simulator.

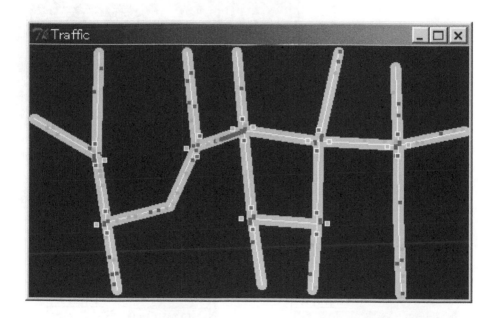

FIGURE 7.6: Simulation of silicon traffic.

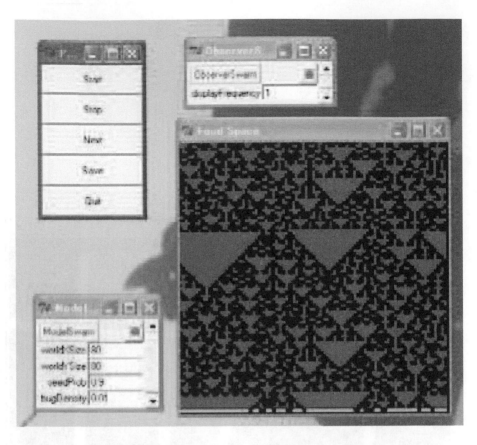

FIGURE 7.8: One-dimensional cellular automaton.

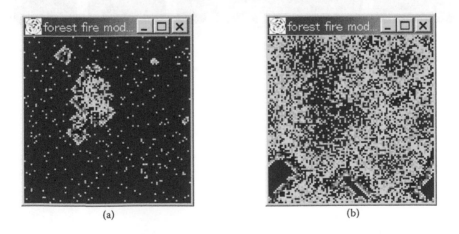

(a)   (b)

FIGURE 7.11: Forest fire examples.

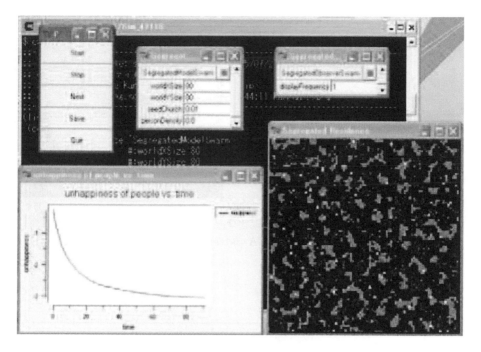

FIGURE 7.13: Schelling's simulation of the segregation model.

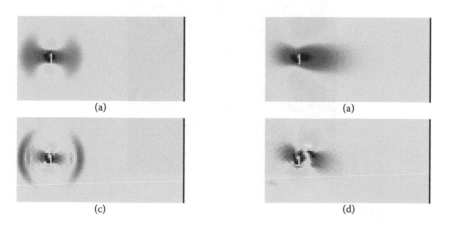

FIGURE 7.18: An example of simulation using the LGA method.

(a) various shells   (b) chambered nautilus (@PNG in 2005)

FIGURE 7.19: CA patterns found on shells.

FIGURE 7.27: Simulating a traffic jam (with SIS).

FIGURE 7.28: Simulating a traffic jam (without SIS).

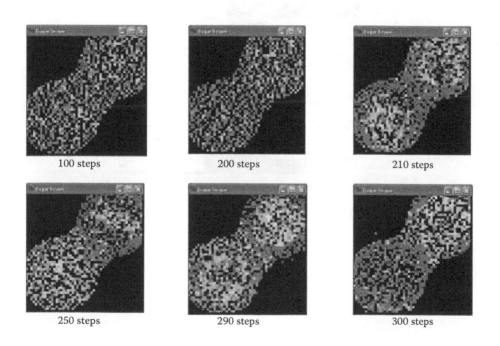

100 steps	200 steps	210 steps
250 steps	290 steps	300 steps

FIGURE 7.48: Combat introduced at the 200th step.

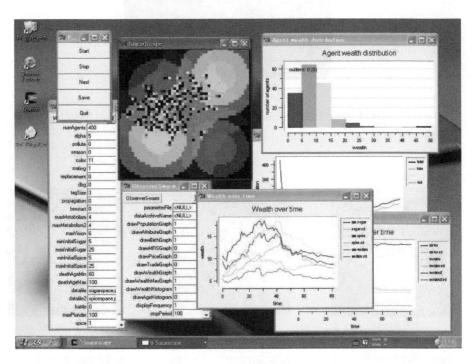

FIGURE 7.57: Sugarscape in Swarm.

Chapter 3

Multi-Agent Simulation Based on Swarm

> Once Zhuangzi dreamt he was a butterfly, a butterfly flitting
> and fluttering around, happy with himself and doing as he
> pleased. He didn't know he was Zhuangzi. Suddenly he woke
> up and there he was, solid and unmistakable Zhuangzi. But
> he didn't know if he was Zhuangzi who had dreamt he was
> a butterfly, or a butterfly dreaming he was Zhuangzi. Be-
> tween Zhuangzi and a butterfly there must be some distinc-
> tion! This is called the Transformation of Things. ("The But-
> terfly Dream," Zhuangzi, B.C. 369?–B.C. 286?, translated by
> Burton Watson)

3.1 Overview of Swarm

In this chapter, we describe the tutorial on how to use Swarm, and aim to implement a practical simulation.[1]

Swarm is a bottom-up model-based simulator which consists of biological agents (such as bugs) and abiotic agents (such as a wall or an obstacle) in an artificial world. The motions of the agents are described by simple rules, and the emergent phenomena due to the interaction between the agents are observed (Fig. 3.1).

The basic characteristics of Swarm are as follows:

1. Swarm is a collection of object-oriented software libraries to assist simulation/programming.

2. Users build simulations by incorporating Swarm library objects in their own programs.

3. Swarm libraries are provided for Java and Objective-C. For the implementation of graphic user interface (GUI), Tcl/Tk has been used.

4. UNIX, Linux, Windows, and Mac OS editions have been published.

---

[1]This tutorial is based on the material by Benedikt Stefansson (UCLA) in Swarmfest 1999 [82, ch. 1]. The original version was implemented in Objective-C.

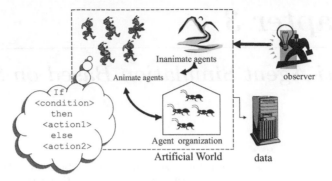

FIGURE 3.1: Bottom-up model.

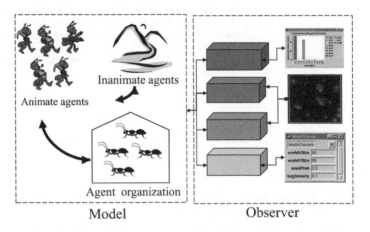

FIGURE 3.2: Swarm modeling.

This chapter focuses mainly on the Java library for the Windows edition. For other versions, refer to the appendix.

Swarm simulation is made up of following two parts (Fig. 3.2):

- `Model`: models the artificial world abstractly, and simulates it.

- `Observer`: observes the simulation of the model and displays it.

The fundamental concept of Swarm is discrete event-driven simulation. In other words, the simulation proceeds in discrete time steps, and further, the interaction between the agents and the simulation procedure has its own schedule of events. Here, let us consider describing the simulation by a usual procedural programming language like C or FORTRAN which can be written as follows:

```
1: get parameters
```

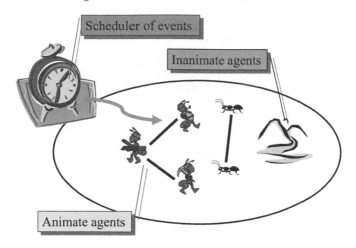

FIGURE 3.3: Object-oriented simulation.

```
2:    initialize  /* initializing data*/
3:    for 1 to timesteps do:
4:        for 1 to num_agents do:
5:            agent-i-do-something
6:        end for
7:        show state /* output the results*/
8:    end for
9: quit
```

In the 5th line, the behavior of the *i*th agent has been described. The user provides the data structure that holds the state of the agent and the action must be implemented. This should be repeated for all agents ("`for`" loop of line 4). By contrast, in Swarm, each agent holds its actions and states itself, and other agents or schedulers act only when called. Thus, the event loop for this purpose is implemented as an object that sends messages to the agent or objects (Fig. 3.3).

Swarm is an object that implements memory allocation and scheduling of events. The basic simulation consists of `Observer` Swarm and `Model` Swarm (Fig. 3.4). Furthermore, Swarm is also being implemented as a virtual machine. In other words, in contrast to normal CPUs that execute instructions, Swarm is the kernel of the virtual CPU that executes `Model` and GUI events (Fig. 3.5).

"`Activity`" is an important concept in Swarm. Multiple Swarm is grouped into a nested structure so that the schedule execution of "`activity`" takes place in a single execution. "`Activity`" object is an abstraction of scheduling events in an artificial world that can be used by the users. This makes it easy to summarize multiple actions and also to arrange the events in the preferred

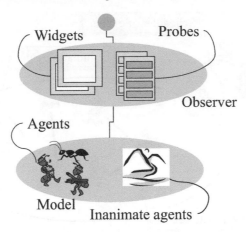

FIGURE 3.4: Model and observer.

order. Since schedule itself is an object, by communicating to other objects, an agent can schedule an event and can also stop any unexpected behavior.

The elements of the "`Activity`" library are as follows (Fig. 3.6):

- `ActivityGroup`: a collection of simultaneous events.

- `Schedule`: (a group of) a collection of sequential events.

- `SwarmActivity`: virtual machine to run schedule.

In Swarm, "`model`" is treated as a structure (Fig. 3.7). "`Schedule`" also merges in accordance with the nested structure and gets executed. In other words, each sub-Swarm schedule merges into the schedule of the Swarm which is one level above. This is repeated recursively. Finally, all the schedules merge with the highest level Swarm.

Memory management is mandatory for dynamic programming that handles large amounts of data such as a multi-agent system. In fact, this is provided in the Swarm library which is available to users. An object is created and annihilated through the idea of a memory zone. "`Collection`" (object or collection) or "sub-Swarm" can be erased by erasing the memory zone.

All the agents and objects in Swarm can be observed. To observe, a probe is pasted on the agent and a message is sent. This way we can change the internal variables of an agent or read their values (Fig. 3.8). In addition, the probe is also used to perform real-time communication between objects through the GUI. The default probemap of the agent displays all the functions and variables (Fig. 3.9). In other words, by using the probe, it is possible to collect information dynamically (at runtime) from one or more agents. Moreover, it is also possible to define a method invocation during the execution of Swarm.

Swarm provides a rich GUI. For example, it has objects to handle line graphs, histograms, raster images, and collection of data. Calculations and

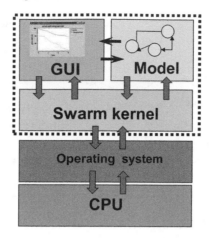

FIGURE 3.5: Swarm as a virtual computer.

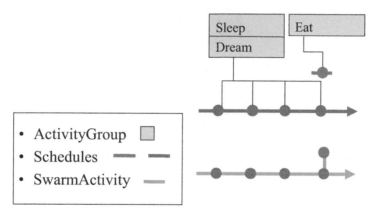

FIGURE 3.6: Elements of Activity library.

updates are carried out through objects and it also provides widgets for the GUI. Displaying averages of data from multiple agents, graph plotting, or plotting using other widgets is done by dynamic access to the dataprobe of an agent (Fig. 3.10). Swarm's "collection" (collection of classes like lists) also supports this.

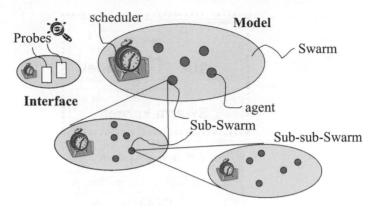

FIGURE 3.7: Recursive structures in Swarm.

FIGURE 3.8: Probes.

3.2   Tutorial

This tutorial is an example of an agent which does random walks and consists of 5 steps shown in Table 3.1.

3.2.1   simpleCBug

The first program is a procedural simulation. After generating a bug agent, it is placed on a two-dimensional plane (of the size worldXSize×worldYSize= 80×80) on the grid. The initial position of the bug is xPos= 40,yPos= 40. Then a random walk is performed 100 times. Let us explain the program briefly. First,

```
import swarm.*;
```

is a kind of charm that imports useful packages of random numbers, etc. Under that is the main function, where the program starts. Main takes the argument

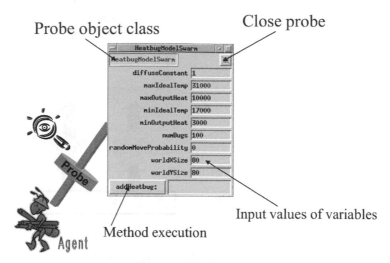

FIGURE 3.9: Probe and GUI.

(args) from the command-line, and the function initSwarm below checks the
args and performs the necessary initialization.

```
Globals.env.initSwarm("bug", "0.2", "foo@nifty.com", args);
```

Here, we specify the "default file" (file named bug.scm) of the "parameter settings." 0.2 is the version name. Other than that, settings such as the address of
the author, etc., are specified. This is detailed in the section simpleSwarmBug3.
    Then, the initial coordinates of the bug are displayed, and function

```
Globals.env.uniformIntRand.getIntegerWithMin$withMax(-1,1);
```

generates the random numbers (integers; -1,0,1), which are added to xPos and
yPos, enabling a random walk.

```
xPos = (xPos + worldXSize) % worldXSize;
yPos = (yPos + worldYSize) % worldYSize;
```

Note that if the bug is outside the field, wrapping becomes necessary (i.e.,
% worldXSize).

3.2.2 simpleObjCBug and simpleObjCBug2

These two are meant to verify the object-oriented programming, and have
no specific significance with regard to Swarm.
    Let us first look at simpleObjCBug. It is the simplest version of the object-
oriented program. Except for the main function, it consists of only one class
and its instance. The "main" function imports the basic library of Swarm,
launches the initSwarm, and performs the memory allocation.
    Class "Bug" is defined in Bug.java:

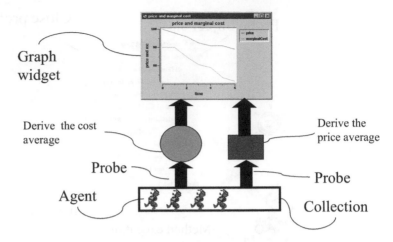

FIGURE 3.10: GUI widgets.

```
public class Bug extends SwarmObjectImpl {
```

This is to declare the class `Bug` as the subclass of `SwarmObjectImpl`. This allows the use of the following function:

- construction: memory allocation

- drop: memory creation and annihilation

In fact, in `Bug.java`, the following definition is the constructor:

```
public Bug(Zone aZone){
     super(aZone);
}
```

`Zone` is the memory space of Swarm. Since "`super`" is the function, it executes the constructor of `SwarmObjectImpl`, which is the super class of `Bug`.

Described in the `main` function,

```
Bug aBug=new Bug(Globals.env.globalZone);
aBug.setX$Y(xPos,yPos);
aBug.setWorldSizeX$Y(worldXSize,worldYSize);
```

are the typical procedures for the generation of an "instance." First, we generate an "instance" of class `Bug` using "`new`," which is then assigned to a variable `aBug`. Memory allocation takes place here. In the next two steps, necessary parameter setting takes place. These methods for the `Bug` are defined in `Bug.java`.

Next, let us move on to `simpleObjCBug2`. In this program, the world where bug walks around and eats the bait is created. This world is defined as

TABLE 3.1: Tutorial contents.

simpleCBug		Procedural programming
simpleObjCBug	1	Object-oriented programming
	2	Bug and FoodSpace interaction
simpleSwarmBug	1	Object responsible for the simulation
		Schedule
	2	Multiple Bugs
	3	Parameter settings with files
simpleObserverBug	1	Display on the 2-dimensional plane
	2	Probe to access the object
simpleExperBug		Multiple models
		Hierarchical schedules
		Graphical display

FoodSpace, and is the subclass of `Discrete2dImp1`. The following lines, inside
FoodSpace.java,

```
public FoodSpace(Zone aZone,int x,int y){
    super(aZone,x,y);
}
```

contain a constructor, in which **super** executes the constructor of
`Discrete2dImp1`.

FoodSpace inherits the internal variables and methods from
`Discrete2dImp1`. It makes the retention and retrieval of values of the two-dimensional space possible. The newly added method by FoodSpace is only
`seedFoodWithProb`. Note that the `getSizeX`, `getDoubleWithMin$withMax`
methods used in this definition are inherited from `Discrete2dImp1`. In
`seedFoodWithProb`, at each point in the FoodSpace a random number between 0 and 1 is generated. If the value of the position is less than `seedProb`,
it changes the value to 1 (that is, `putValue$atX$Y(1,x,y)`). This is to
achieve the bait at concentration of `seedProb`. In this program, the variable
(`foodSpace`) of `FoodSpace` is defined as an internal variable of `Bug`. It holds
the world in which it moves. In the part below the method "**step**," if the new
location has bait, it eats the bait (substitutes the value of 0) and displays that
position.

```
if (foodSpace.getValueAtX$Y(xPos,yPos) == 1){
    foodSpace.putValue$atX$Y(0,xPos,yPos);
    System.out.println("I found food at X = " + xPos + " Y = "
                              + yPos +"!");
}
```

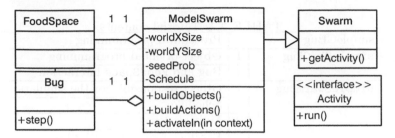

FIGURE 3.11: Class hierarchy in simpleSwarmBug.

3.2.3   simpleSwarmBug

The typicality of Swarm can be shown with this program. In the simpleSwarmBug of the third phase, the object ModelSwarm is introduced (Fig. 3.11). It manages all the models (object of Bug and FoodSpace, etc., parameters such as the size of the world, schedule, etc.). Often, Model becomes the nested structure of multiple swarms.

The generation of Swarm is done by the following steps:

1. construction: initialization of memory and parameters

2. buildObjects: construction of all agents and objects in the model

3. buildActions: sequence of events and definition of schedule

4. activate: beginning of the execution and merging into the top level swarm

The above steps correspond to the following lines in the simpleSwarmBug.java:

```
observerSwarm=new ObserverSwarm(Globals.env.globalZone);
observerSwarm.buildObjects();
observerSwarm.buildActions();
observerSwarm.activateIn(null);
```

The details of these will be explained below.

3.2.3.1   Initialization

The constructor of ModelSwarm.java is as follows:

```
1: public ModelSwarm(Zone aZone){
2:   super(aZone);
3:
4:   worldXSize = 80;
5:   worldYSize = 80;
6:   seedProb   = 0.5;
```

```
7:  bugDensity = 0.1;
8: }
```

The "super" in the second line executes the constructor of the super class "SwarmImp" and allocates the memory. In lines 4–7, initial values are being assigned. Its own object (this) can be omitted here. More precisely, this should be written as follows:

```
4: this.worldXSize = 80;
4: this.worldXSize = 80;
5: this.worldYSize = 80;
6: this.seedProb   = 0.5;
7: this.bugDensity = 0.1;
```

Here, let us explain the "aZone" in the first line. In Swarm, all the objects are created in the memory region called "zone." Methods for allocating and freeing the memory for that purpose are provided. Memory for "instance" and it's internal variables are also supplied. Moreover, "zone" keeps track of all the objects created there, and memory can be re-used simply by dropping the "zone." In other words, a signal to all objects can be sent for self-destruction.

The three main parts of "zone" are summarized as follows:

1. The initSwarm(...) function inside the "main," executes various functions that generate global memory areas.

2. In modelSwarm=newModelSwarm(Globals.env.globalZone) inside the "main" function, instances are created.

3. ModelSwarm class' constructor modelSwarm() is executed.

3.2.3.2   Construction of Agents

This is defined in buildObjects of ModelSwarm.java. The purpose of this method is to generate the instance of the class required for the representation of objects in simulation, and to set the parameters in them. As we can see from the definition of setFoodSpace, it is an advantage of object-oriented programming that the functions for the settings of variables are ready through inheritance.

3.2.3.3   Construction of Schedules

A schedule is constructed by the following method called buildActions in ModelSwarm.java.

```
1: public Object buildActions(){
2:   modelSchedule=new ScheduleImpl(this,1);
3:   try {
4:       modelSchedule.at$createActionTo$message(0,aBug,
```

```
5:      new Selector(Class.forName("Bug"),"step",false));
6:  } catch (Exception e) {
7:      System.out.println ("Exception: " + e.getMessage ());
8:      System.exit(1);
9:  }
10:
11: return this;
12: }
```

In the second line, the instance of the class "Schedule" is generated. The second argument "1," means "Execute this schedule at each time step." If we write the natural number n here, that schedule will be executed every n time steps. In lines 4–5, an event is assigned to the schedule. Here, in the instance "aBug" of class "Bug," "Action" is generated which executes the method called "step." The first argument indicates the time of first boot. In other words, in this case it executes at time step 0. Lines 6–9 are for handling errors if the occurred. That means, by sending at$createActionTo$message (start, instance, method) method to the instance modelSchedule of schedule, Action is generated. Here, the first argument "start" is the starting time of "Action," and an event called Action is set which sends "method" to "instance."

Furthermore, the method Selector(Class, string, boolean) retrieves the method defined by the string-name "string" of the class "Class." The third argument "boolean" is false when it retrieves the methods of Java and true in the case of Objective-C.

3.2.3.4   Activation of Swarm

In the "main" function, by calling

```
modelSwarm.activateIn(null);
```

a schedule gets activated. Since there is only one Swarm in simpleSwarmBug, the argument becomes "null." Specifically, the following method is called:

```
public Activity activateIn(Swarm context){
    super.activateIn (context);
    modelSchedule.activateIn(this);
    return getActivity ();
}
```

Merge and activation of schedule are initiated by this method.

3.2.3.5   State of Execution

As the program starts, first the "ModelSwarm" is executed, and ModelSwarm generates Bug and FoodSpace (method: buildObjects).

In the body of the simulation, Bug repeats the method "step." This is

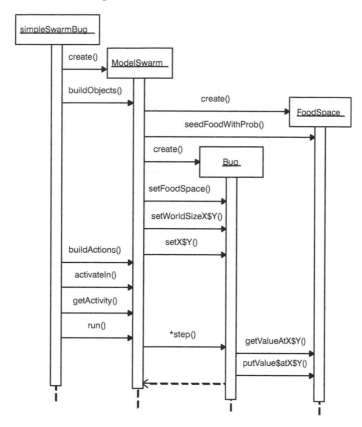

FIGURE 3.12: Execution sequence in simpleSwarmBug.

implemented (in the method "`buildActions`") by registering "`action`" in the schedule: "Send the "`step`" method execution message to the "`Bug`" instance" while "starting at time 0 and repeating with a time interval of 1." Schedule is activated by the method "`activateIn`," and is executed by "`run`." The state of execution is shown in Fig. 3.12.

3.2.4 simpleSwarmBug2

Multiple agents are handled in this program. Since multiple bugs are handled as a set, the instance "`bugList`" of class "`List`" is used (Fig. 3.13). The concentration of bugs is set by the variable "`bugDensity`."

```
if (Globals.env.uniformDblRand.getDoubleWithMin$withMax(0.0,1.0)
                                    < bugDensity){
```

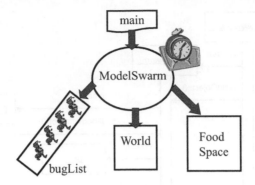

FIGURE 3.13: simpleSwarmBug2.

contains that creation part, in which random variables from 0 to 1 are generated. If that value is smaller than the "bugDensity," then an instance of Bug is created and initialized as shown below:

```
aBug=new Bug(this);
aBug.setWorld$Food(world,food);
aBug.setX$Y(x,y);
```

In the following line:

```
bugList.addLast(aBug);
```

a newly created instance is being added at the back of bugList. Moreover,

```
reportBug=(Bug)bugList.removeFirst();
bugList.addFirst(reportBug);
```

substitutes the variable "reportBug" at the front of the list. As will be mentioned in a later section, reportBug is used for outputting the position.

At the same time, multiple "action" messages are organized into one for activation, and then implemented as the class instance of "ActionGroupImp1," which is the implementation of the "ActionGroup" interface. This is useful for denoting the concurrent execution for multiple bugs. For this purpose, we need to declare the variable modelActions in

```
ActionGroup modelActions;
```

and then, using the buildActions method, the following instance is created:

```
modelActions=new ActionGroupImpl(this);
```

After that, in the following lines:

```
modelActions.createActionForEach$message(bugList,
    new Selector(Class.forName("Bug"),"step",false));
                modelActions.createActionTo$message(reportBug,
    new Selector(Class.forName("Bug"),"report",false));
```

multiple actions are defined. Here, through `createActionForeach$message`, "action" executes for all the elements of `bugList`. Moreover, methods (`report`) for other instances (`reportBug`) are also added.

Next, schedule is created with the following lines:

```
modelSchedule=new ScheduleImpl(this,1);
modelSchedule.at$createAction(0,modelActions);
```

Here, the frequency of schedule execution is 1 time step, starting at time 0. To ensure that there is only one bug in a particular location, "world" is created as the instance of class "Grid2d." For the method "buildObjects" inside `ModelSwarm.java`, creation and initialization are done by the following lines (each is filled with "null"):

```
world=new Grid2dImpl(this,worldXSize,worldYSize);
world.fillWithObject(null);
```

For the method "step" in "Bug," using the following lines:

```
if (world.getObjectAtX$Y(newX,newY) == null){
    world.putObject$atX$Y(null,xPos,yPos);
    xPos = newX;
    yPos = newY;
    world.putObject$atX$Y(this,newX,newY);
}
```

the past position (xPos,yPos) is cleared (`null`), and the instance of "Bug" is described in a new location (newX,newY). In the following "if" loop,

```
if (foodSpace.getValueAtX$Y(xPos,yPos) == 1){
    foodSpace.putValue$atX$Y(0,xPos,yPos);
    haveEaten=1;
}
```

if there is a bait in the new location, the bug eats it (resets to 0), and makes the "haveEaten" flag 1. This flag is used to determine the display in the method "report."

3.2.5   simpleSwarmBug3

This is not much different from "simpleSwarmBug3." The only difference is that it reads the default parameters from an external file for initialization. The description for this is in the following line of the "main" function:

```
Globals.env.initSwarm("bug", "0.2", "foo@nifty.com", args);
```

Here, the setting of the default file of parameter settings is done. The extension "scm" added to the first argument becomes the filename (that is, `bug.scm`). The conetents of this file are written as follows:

```
(list
 (cons 'modelSwarm
   (make-instance 'ModelSwarm
                   #:worldXSize 80
                   #:worldYSize 80
                   #:seedProb 0.9
                   #:bugDensity 0.01)))
```

This specifies the initial values of parameters using the description of "Scheme" (one of the dialects of the "Lisp" language). By changing this number, we can set the initialization parameters without having to recompile.

Furthermore, in this program, the "ModelSwarm" method inside "ModelSwarm.java" also sets the initialization parameters as follows:

```
worldXSize = 80;
worldYSize = 80;
seedProb = 0.8;
bugDensity=0.1;
```

However, this setting is given priority over the one described in the "scm" file.

3.2.6   simpleObserverBug

In `simpleObserverBug`, we can learn how to use the GUI. Perhaps the biggest advantage of "swarm" is that the GUI can be easily constructed. The possible GUI features are summarized as follows:

- Provides interactive access (`probe`) to the object.

- Displays the state of distribution of objects and numerical values over the two-dimensional plane.

- Displays line graphs and histograms.

Although "Object2dDisplay" handles the monitor display, "ZoomRaster" is responsible for the actual display (window). `ZoomRaster` displays the data from the grid (Fig. 3.14).

This program displays the distribution of bait from "FoodSpace" and makes the bugs display their positions. For this purpose it uses the following three classes (Fig. 3.15):

- `ColorMap`: associates the numbers and colors on the palette.

FIGURE 3.14: Display of object distribution (ZoomRaster).

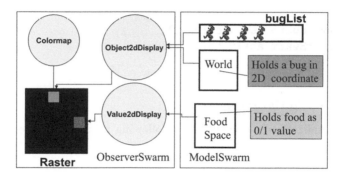

FIGURE 3.15: Display with Swarm.

- Value2dDisplay: maps the x- and y-grid data on the raster.

- Object2dDisplay: processes the mouse clicks and the data display from the agents.

The generation of the "ObserverSwarm" is done in the following steps (Fig. 3.16):

1. construction: initializes memory and parameters.

2. buildObjects: constructs ModelSwarm, i.e., graph, raster, and probe.

3. buildActions: defines the schedules and order of events for GUI.

4. activate: merges into the top-level "swarm" and starts "swarm" execution.

All these are equivalent to the following lines in simpleSwarmBug.java.

```
Globals.env.initSwarm("bug", "0.2", "foo@nifty.com", args);
observerSwarm=new ObserverSwarm(Globals.env.globalZone);
observerSwarm.buildObjects();
```

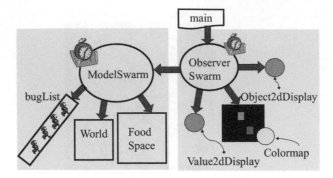

FIGURE 3.16: simpleObserverBug.

```
observerSwarm.buildActions();
observerSwarm.activateIn(null);
```

They will be explained in detail below.

3.2.6.1   Initialization

Initialization is done with the following method in `ObserverSwarm.java`:

```
public ObserverSwarm(Zone aZone){
    super(aZone);
    displayFrequency=1;
}
```

where the memory is allocated and the display frequency is set.

3.2.6.2   Generation of object

By using the "buildObjects" method inside `ObserverSwarm.java`, the generation of `ModelSwarm` is done in the following way:

```
super.buildObjects();
modelSwarm = new ModelSwarm(this);
getControlPanel().setStateStopped();
modelSwarm.buildObjects();
```

The message "`getControlPanel`" is the command which waits for the simulation to start until the "`Start`" button is pressed.

After this, processing is performed as follows:

- `colorMap`: it allocates the numbers to the display colors. Specifically, it allocates 0 to black (blank), 1 to red (bait), 2 to green (bug).

- Generation of `worldRaster`: displays a two-dimensional grid. "`title`" and "`colorMap`" are allocated for that purpose.

- Generation of foodDisplay: foodDisplay is generated as an instance of "value2dDisplayImp1" and bait (red) is drawn using the inheritance feature of this class. The fourth argument "modelSwarm.getFood" is defined inside "modelSwarm" and returns "food," which is an instance of FoodSapce.

- Generation of bugDisplay: bugDisplay is generated as an instance of "object2dDisplayImp1" and a bug (green) is drawn using the inheritance feature of this class. "drawSelfOn," included in the argument, is the method defined at the end of Bug.java, and draws the insects with the value of 2 (i.e., green).

3.2.6.3 Construction of scheduling

By using the method "buildActions" inside ObserverSwarm.java:, a schedule is constructed as follows:

1. Generation of actions for modelSwarm: modelSwarm.buildActions();

2. Generation of ActionGroup: displayActions=new ActionGroupImp1 (this);
   - action for foodDisplay: execution of display
   - action for bugDisplay: execution of display
   - action for worldRaster: execution of drawSelf
   - action for updating the GUI: execution of doTkEvents

3. Generation of displaySchedule: schedule with the frequency of displayFrequency, taking the starting time as 0.

3.2.6.4 Activation of Swarm

By calling

observerSwarm.activateIn(null);

in the "main" function, the schedule gets activated. Specifically, the method below is being called:

```
public Activity activateIn(Swarm context){
    super.activateIn(context);
    displaySchedule.activateIn(this);
    return getActivity();
}
```

Swarm has a hierarchical structure as shown in Fig. 3.17. In the top level Swarm, if the activation starts by calling activateIn(null), the activation gets transmitted to the lower levels one after the other. This can start the merging and activation of the schedule. The "main" function generates ObserverSwarm, and the ObserverSwarm generates ModelSwarm in its own memory as a sub-swarm. Moreover, ModelSwarm generates an agent and activates itself inside ObserverSwarm.

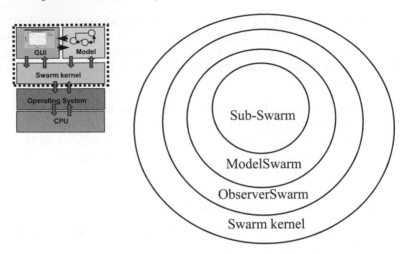

FIGURE 3.17: Swarm hierarchy.

3.2.7   simpleObserverBug2

In "`simpleObserverBug2`," the probe and two-dimensional plane display are implemented (Fig. 3.18). Through the use of the probe, it becomes easy to access the field of an object or to call methods. The probe has the types mentioned below:

- Object and interface
  - `VarProbe`: probes interface variables
  - `MessageProbe`: probes methods
- Generation of GUI for objects:
  - `ProbeMap`: collective use of `VarProbe` and `MessageProbe`

An example is shown in Fig. 3.19. Steps to generate a probe are given below (see Fig. 3.20):

1. Generate an `EmptyProbeMap` instance.

2. Make the GUI usable by pasting `VarProbe` or `MessageProbe` in the variable and message.

3. Put each probe in `ProbeMap`.

4. Request the generation of a real widget from `probeDisplayManager`.

Let us look at a particular program of `simpleObserverBug2`. First, the following description is given in the constructor of `ObserverSwarm`:

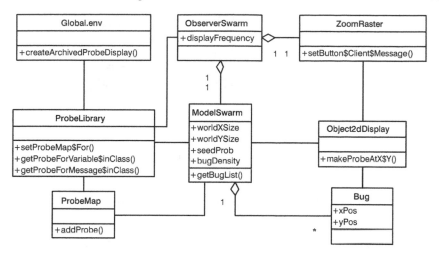

FIGURE 3.18: Class hierarchy in simpleObserverBug2.

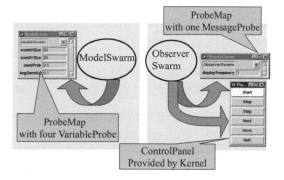

FIGURE 3.19: Probes in simpleObserverBug2.

```
1: probeMap=new EmptyProbeMapImpl(aZone,this.getClass());
2: probeMap.addProbe(Globals.env.probeLibrary.
     getProbeForVariable$inClass("displayFrequency",this.getClass()));
3: Globals.env.probeLibrary.setProbeMap$For(probeMap,
     this.getClass());
```

In the first line, the instance of the implemented class "`emptyProbeMapEmp1`" is being generated for the interface "`emptyProbeMap`." In the second line, in order to make the internal variable "`displayFrequency`" displayable, `VarProbe` is pasted to this probe. The third line does the setting of the probe as "`probeMap`."

Again, inside the constructor of `ModelSwarm`, a different probe is generated as follows:

```
probeMap=new EmptyProbeMapImpl(aZone,this.getClass());
```

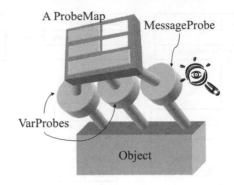

FIGURE 3.20: Objects and probes.

```
probeMap.addProbe(Globals.env.probeLibrary.getProbeForVariable$inClass
    ("worldXSize",this.getClass()));
probeMap.addProbe(Globals.env.probeLibrary.getProbeForVariable$inClass
    ("worldYSize",this.getClass()));
probeMap.addProbe(Globals.env.probeLibrary.getProbeForVariable$inClass
    ("seedProb",this.getClass()));
probeMap.addProbe(Globals.env.probeLibrary.getProbeForVariable$inClass
    ("bugDensity",this.getClass()));
Globals.env.probeLibrary.setProbeMap$For(probeMap,this.getClass());
```

"varProbe" is set, which displays the the internal variables, i.e., `worldXSize`, `worldYSize`, `seedProb`, and `bugDensity`.

After that, in "`buildObjects`" inside `ObserverSwarm.java`, the following lines request `probeDisplayManager` for the generation of a specific "widget":

```
Globals.env.createArchivedProbeDisplay (modelSwarm,
                                          "modelSwarm");
Globals.env.createArchivedProbeDisplay (this, "observerSwarm");
```

Here, "`this`" is `observerSwarm` itself. The second argument's string is the title to be displayed when displaying the probe. The next line shows the instruction to wait to start the execution in response to pressing the "`start`" button.

```
getControlPanel().setStateStopped();
```

The execution screen is shown in Fig. 3.21. As can be seen in the figure, four windows are generated. These are

- Control button (Start, Stop, etc.)

- Fields (bug and bait) that `worldRaster` displays

- `ModelSwarm`'s variable display window

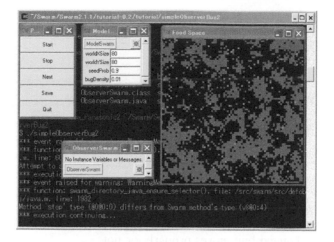

FIGURE 3.21: simpleObserverBug2.

- `ObserverSwarm`'s variable display window

Out of these, the last two windows are the ones generated under the probe.

If some object is displayed, that object's probe can be called by clicking it. By "setButton$Client$Message" of "ZoomRaster," it is possible to specify an action taken when the mouse is clicked. For example, let us have a look at the following definition of `buildObjects` inside the `ObserverSwarm.java`:

```
worldRaster.setButton$Client$Message(
  3,bugDisplay,new Selector(bugDisplay.getClass(),
                            "makeProbeAtX$Y",true));
```

Here, the method "makeProbeAtX$Y" of `Object2dDisplay` is called, and the probe of the object (instance of `Bug`) is generated as shown in Fig. 3.22(a). By executing this method, if you right-click "bug," you can see the variable value. However, for that purpose, the internal variables inside `Bug.java`,

```
int xPos, yPos;
```

have to be changed to "`public`" declaration as follows:

```
public int xPos, yPos;
```

Let us confirm this feature by rewriting and then recompiling (note that "`*.class`" files should be deleted before recompiling). After pressing "`stop`," you can right-click on the appropriately chosen bug. By doing that, the coordinates (`xPos`, `yPos`) of the bug will be displayed on the window (Fig. 3.22(a)).

Again, let us try to change the coordinates of the bug, which are displayed on pressing "`stop`" (do not forget to press "`enter`" at this point). Next, one

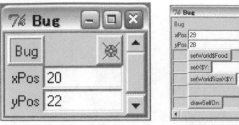

(a) Default probe   (b) Complete probe

FIGURE 3.22: Probes for Bug.

step at a time is executed every time "**step**" is pressed. Now you can confirm whether the changed bug moves properly or not.

The first argument of `worldRaster.setButton$Client$Message` contains the following meanings:

- 1. display on left-click

- 2. display on middle-click

- 3. display on right-click

The third argument of the function "**selector**" is true since the retrieving method has been described in Objective-C.

Fig. 3.22(a) is the probe for Bug, and public methods can be registered if specified. By default, only the fields are registered. If you right-click the class name on the top left, all public fields and methods will be displayed (Fig. 3.22(b)). In addition, if you click the button on the top right, all public fields and methods of the superclass will be displayed. Therefore, the probe is a powerful tool when working with a simulator.

3.2.8   simpleExperBug

"`simpleExperBug`" repeats the execution of models and displays those statistics as a line graph.

It boots the `ExperSwarm` from the "main" function. Then, in the activation of `ObserveSwarm`, `ModelSwarm` goes in sequence (Fig. 3.23). Here we use Swarm itself to control the repeated execution of `ModelSwarm`. The number of iterations of the experiments is retained in the internal variable called "numModelsRun." In "experSwarm," the execution is made as follows:

1. "numModelsRun" is initialized to be 0 by the constructor of `ExperSwarm`. It also provides `VarProbe` for the `numModelsRun`.

2. `EZGraph` is provided to display line graphs for the results inside "`buildObjects`" methods.

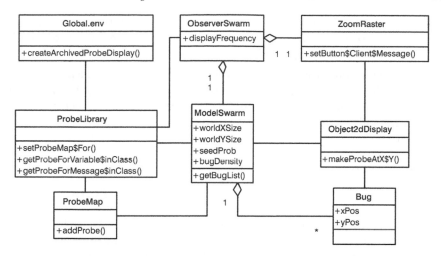

FIGURE 3.23: Execution sequence in simpleExperBug.

3. The action group "`experActions`" is generated which is made inside "`buildActions`" and is set as a schedule, with the starting time as 0 and repeating every one time step.

- `buildModel`: Generates "`varProbe`"(modelProbeMap) of internal variables only at the initial execution (when `numModelsRun` is 0). Also sends the `initializeModel` method to `parameterManager` and sets the parameters (parameters are updated at `checkToStop`). Moreover, it executes `buildObjects`, `buildActions`, `activateIn` for `modelSwarm` so as to construct the `modelSwarm`.

- `RunModel`: Sends the "run" method to `modelSwarm` and executes once. After that, it increments the number of executions after displaying (`numModelsRun++`).

- `doStats`: Displays the execution time of models. Execution time is obtained by sending the method called "`setTime`" to `modelSwarm` (described later).

- `showStats`: Performs the display of results.

- `logResults`: Displays the log of results (this part is not yet implemented).

- `dropModel`: Frees the memory of `modelSwarm`.

- `checkToStop`: Sends the `stepParameters` method to `parameterManager` and changes the parameters. As defined in `ParameterManager.java`, this increases the value of `seedProb` and `bugDensity`, by the specified value (`seedProbInc`, `bugDensityInc`). In case the value exceeds the maximum set value (`seedProbMax`, `bugDensityMax`), "null" is returned by the execution of the "stepParameters" method and the execution ends. In other words, it depends on the execution of getControlPanel(). setStateStopped().

Whenever modelSwarm is constructed by the buildObjects method of modelSwarm.java, the "time" variable is initialized to 0. In the next buildActions method, the "modelSwarm" action group is generated consisting of the following actions, and at starting time 0, it is set as a schedule of 1 time step.

- Execution of the "step" method for all the "bug" objects of bugList

- Execution of the "checkToStop" method for modelSwarm

In the "checkToStop" method, it terminates if no bait is left, else the time is incremented (time++). The "setTime" method of modelSwarm only returns the variable "time." Therefore, by sending the method "setTime" to modelSwarm at the execution termination time, you can obtain the time steps elapsed until the execution termination for that iteration.

The object that displays the line graph in this way is "EZGraph." EZGraph is a wrapper of some of the objects, and is set by following steps:

1. Generation of instance

2. Settings of title and axis labels

3. Settings of display series

These correspond to the following parts of buildObjects inExperSwarm.java:

```
1: resultGraph=new EZGraphImpl(
1:     this,"Model Run Times","Model #","Run Time","resultGraph");

2: resultGraph.enableDestroyNotification$notificationMethod
2:    (this, new Selector(getClass(),"_resultGraphDeath_",false));

3: resultGraph.createSequence$withFeedFrom$andSelector(
3:    "runTime",this,new Selector(this.getClass(),"getModelTime",
       false));
```

The value of the time variable obtained by the method "setModelTime" is successively displayed as the value of the y-coordinate. "runTime" in the last line is an example name of this line graph.

The execution screen when ExperSwarm is executed is shown in Fig. 3.24. Note that, in the declaration of class Bug of Bug.java, the "final" represents the class that cannot be extended (cannot be inherited).

```
public final class Bug extends SwarmObjectImpl {
```

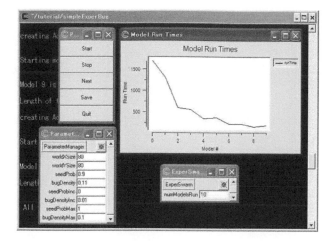

FIGURE 3.24: ExperSwarm.

The method "`setActionCache`" is being used for making the scheduler as a thread. For more information, see the manual.

In `public Object buildObjects()` of `ModelSwarm.java`, random number seeds can be provided as follows:

`Globals.env.randomGenerator.setStateFromSeed(randomSeed);`

where `randomSeed` is an integer value. Using the same seed can reproduce specific simulation results. If we use a time variable in a seed, every time we boot, simulations with different behaviors can be observed.

Chapter 4

Evolutionary Simulation

> But we really know little about the minds of the lower animals. It cannot be supposed, for instance, that male birds of paradise or peacocks should take such pains in erecting, spreading, and vibrating their beautiful plumes before the females for no purpose. We should remember the fact given on excellent authority in a former chapter, that several peahens, when debarred from an admired male, remained widows during a whole season rather than pair with another bird (Charles Robert Darwin [23, p. 490]).

4.1 Simulation of sexual selection

Why does the peacock have such wastefully beautiful feathers?

Why did the zebra develop such a striped pattern?

Biologists are divided in their opinions regarding the excessive beauty of certain animals. Does it fulfill a function of distinguishing individuals, of promoting mutual grooming, of protecting the body from harmful insects, or of adjusting the body temperature? Perhaps the aesthetics are a handicap?

Researchers conceive that such wonderful patterns might be unsuitable for adaptation. The maintenance of beauty requires energy (cost) and prevents the animal from hiding effectively when discovered by predators. The feathers of a peacock do not provide an advantage when running away from a threat.

Even Darwin was troubled by this problem when building arguments for his theory of evolution. Why did evolution produce certain traits (behavioral and physical characteristics) that are seemingly detrimental to individuals? In this chapter, we present several hypotheses explaining this, together with verification experiments based on genetic algorithms (GAs).

4.1.1 Sexual selection in relation to markers, handicaps, and parasites

In the following paragraphs, let us consider the difference in opinion between Darwin and Wallace in regard to the striped pattern of the zebra.

The zebra is conspicuously striped, and stripes on the open plains of South Africa cannot afford any protection. Here we have no evidence of sexual selection, as throughout the whole group of the Equidae the sexes are identical in colour. Nevertheless he who attributes the white and dark vertical stripes on the flanks of various antelopes to sexual selection, will probably extend the same view to the Royal Tiger and beautiful Zebra. [23, p. 302]

It may be thought that such extremely conspicuous markings as those of the zebra would be a great danger in a country abounding with lions, leopards, and other beasts of prey; but it is not so. Zebras usually go in bands, and are so swift and wary that they are in little danger during the day. It is in the evening, or on moonlight nights, when they go to drink, that they are chiefly exposed to attack; and Mr. Francis Galton, who has studied these animals in their native haunts, assures me, that in twilight they are not at all conspicuous, the stripes of white and black so merging together into a gray tint that it is very difficult to see them at a little distance. [123, p. 220]

Being the subject of "a delicate arrangement [13]," Wallace is famous in a paradoxical sense (one of his achievements in biology was the proposal of the Wallace boundary found in biota). He and Darwin independently and almost simultaneously devised the theory of evolution by natural selection, and the paper sent by Wallace from the Malay Archipelago surprised the slow writer Darwin. Eventually, both papers were presented simultaneously in 1858 at the Linnean Society after Darwin had made "minor adjustments" in relevant parts of his paper.

In On the Origin of Species, Darwin argues that a female chooses a mate after seeing a display of marvelous plumage and unusual, outlandish body movements [21A]. Being chosen by the opposite sex in this way is known as "elimination through sexual selection" or simply "sexual selection."

The law of battle for the possession of the female appears to prevail throughout the whole great class of mammals. Most naturalists will admit that the greater size, strength, courage, and pugnacity of the male, his special weapons of offence, as well as his special means of defence, have been acquired or modified through that form of selection which I have called sexual. This does not depend on any superiority in the general struggle for life, but on certain individuals of one sex, generally the male, being successful in conquering other males, and leaving a larger number of offspring to inherit their superiority than do the less successful males. [23, ch. XVIII]

Fisher supported the theory of sexual selection proposed by Darwin (Fig. 4.1). In a seminal work entitled The Genetical Theory of Natural Se-

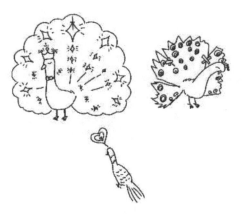

FIGURE 4.1: The sexual selection.

lection, Fisher presented an adequate explanation of the female animal preferences described by Darwin in his theory [36A]. The premise of Fisher's theory is that successful achievement generates further successful achievement, resulting in positive feedback with the potential for explosive growth. For example, if the preference for a long tail becomes increasingly successful, the tails of males in later generations will become longer, and females will prefer the males with even longer tails. This results in the development of yet longer tails and a preference for long tails in females. Since success depends on the rate of occurrence, the process progresses in a self-reinforcing manner.

The optimal strategy is that practiced by the majority, and mimicking the actions adopted by most individuals is thus preferable. Therefore, selection based on a long tail and the preference for a long tail progress together. The appearance of males and the preference in females for a certain appearance evolve hand-in-hand in such a way that they reinforce each other. This process is considered to have eventually given rise to the extraordinary tail feathers of the peacock, and explains why the evolution and preference of a certain appearance often exceeds a moderate level and becomes a runaway process [21, p. 284]. Such rapid growth is referred to as "the runaway effect" by Fisher.

In contrast, Wallace adopted the point of view that coloration is ultimately derived from "protective colors and marks." This idea about protective colors explains both concealing coloration (which hides the body by deceiving the eyes of an adversary) and conspicuous coloration. In other words, Wallace points out that although it might appear at first sight that the coloration of some animals is extremely noticeable, this coloration is in fact concealing in the natural habitat of the animal. Examples supporting this view include the zebra, the tiger, and the giraffe, which blend seamlessly into their respective native environments. In addition, inedible animals that use bright coloration as a warning, and animals that mimic such coloration, are in excellent agreement with this explanation. Furthermore, colored markings distinguish individuals

FIGURE 4.2: The handicap principle.

of the same species, which serves as an aid for social animals (such as ants and bees) to stay together or to recognize potential partners. The bright coloration of various birds can also be explained in this way [21, pp. 176–178].

Although all the details about coloration as proposed by Wallace do not necessarily hold true, and despite the large number of demonstrated counterexamples, the general outline of Wallace's argument is well constructed, and has been adopted as a fundamental viewpoint in Darwinian theory.

Conversely, the "handicap principle" proposed by the Israeli biologist Amotz Zahavi accounts for the seemingly paradoxical existence of animals that pay an extraordinarily high cost for their appearance. With its extravagant feathers, the peacock makes a statement that it can afford its beautiful appearance, since its nourishment is plentiful and its agility and stamina eliminate the risk of being discovered by predators. Thus, the peacock is in a superior position even with the disadvantage imposed by this handicap (Fig. 4.2). In the same manner, bucks with extraordinary horns make a statement that despite their handicap, their nourishment is sufficient and they are strong enough to win a fight.

> Why does the gazelle reveal itself to a predator that might not otherwise spot it? Why does it waste time and energy jumping up and down (stotting) instead of running away as fast as it can? The gazelle is signaling to the predator that it has seen it; by "wasting" time and by jumping high in the air rather than bounding away, it demonstrates in a reliable way that it is able to outrun the wolf. [130, pp. xiii–xiv]

A third theory that emphasizes the role of parasites was recently proposed, sparking controversy. This theory was developed by the animal behaviorist Hamilton, and resembles the handicap principle. Using the male's feathers as a clue, a female assumes that she is choosing a partner with superior genes. A male's genes determine its immunological resistance to parasitic agents (bac-

FIGURE 4.3: The parasite principle.

teria, viruses, and parasites), since maintaining large beautiful feathers is difficult if the body has been severely damaged by parasites. Therefore, choosing a male with spectacular feathers equals choosing that with strong resistance to parasites (and hence strong capability; Fig. 4.3). Although, up to this point, Hamilton's theory shares the same line of thought as the handicap theory, it differs drastically from the latter in the following sense. While predators in Zahavi's theory are more or less known, new parasites constantly emerge, and the parasites themselves are extremely diverse. Therefore, a considerably stronger resilience must be maintained than that necessary for avoiding predators. The magnificence of a male's feathers clearly shows its resilience against current parasites, and the female wishes to obtain this information by all means possible. Hence, the feathers of the peacock have acquired this extravagance, which also appeals to the female.

To verify his theory, Hamilton investigated several species of birds (Passeriformes living in North America) in terms of their resistance to parasites, by ranking on a six-level scale the plumage and the song complexity of males. The results were subsequently compared with research data on the concentration of trematodes and protozoans in the blood of these birds, and a correlation with assigned ranks was investigated. Unsurprisingly, species with more extravagant feathers and more complex songs also had greater resilience to parasites. Investigating the feathers of the females, the correlation was almost indistinguishable from that observed in males. Hamilton's parasite hypothesis was thus considered to be verified. However, a number of problems with this experiment have been pointed out, and the debate continues.

We now turn our attention to the use of GAs in sexual selection research.

4.1.2 The Kirkpatrick model

Mark Kirkpatrick [69] studied a mathematical model of sexual selection based on population genetics. In this model, each individual carries two types of genes denoted T (traits, features) and P (preferences), and each gene has a pair of alleles (T_0, T_1 and P_0, P_1, respectively). P encodes the preference of a female with respect to trait T of a male such that females carrying P_i

prefer males carrying T_i $(i = 0, 1)$. These preferences depend on the rate of occurrence, and the probability of a P_i female breeding with a T_j male is set as

$$P(T_0 \mid P_0) = \frac{a_0 t_0'}{t_1' + a_0 t_0'}, \tag{4.1}$$

$$P(T_1 \mid P_1) = \frac{a_1 t_1'}{t_0' + a_1 t_1'}, \tag{4.2}$$

$$P(T_1 \mid P_0) = 1 - P(T_0 \mid P_0), \tag{4.3}$$

$$P(T_0 \mid P_1) = 1 - P(T_1 \mid P_1). \tag{4.4}$$

Here, t_k' denotes the rate of occurrence of mature T_k males.

Let us assume that trait T_1 is detrimental to survival. This precisely reflects the scenario that predators can easily discover peacocks with beautiful feathers, which are preferred by peahens. We therefore consider it more difficult for males carrying T_1 to survive than males carrying T_0. Specifically, we take the probability of survival of T_1 males as $1 - s$ times that for T_0 males (where $0 < s < 1$ is a real number). Although males also carry P, this gene is not expressed in their appearance, and similarly for T in females.

In the Kirkpatrick model, the initial population has the same number of males and females. In each successive generation, a proportion of T_1 males die before procreating; explicitly, s times the number of T_1 males are killed. Each female then chooses a male for mating, where P_i females prefer T_i males following Eqs. 4.1 and 4.2. After breeding, the next generation is produced, and this process is repeated for a certain number of generations.

Kirkpatrick performed a simulation without devising an actual version of this system. As a result, Kirkpatrick derived an equation where the rate of occurrence of P_1 females and T_1 males is balanced (a state of equilibrium, where no further changes occur). Here, taking p_1 and t_1 as the respective occurrence rates of P_1 and T_1, Kirkpatrick presented the following equilibrium equation.

$$t_1 = \begin{cases} 0 & p_1 \leq V_1, \\ \frac{(a_0 a_1 - 1)(1-s)}{(a_0 + s - 1)[a_1(1-s)-1]} p_1 - \frac{1}{a_1(1-s)-1} & \\ & V_1 < p_1 < V_2, \\ 1 & V_2 \leq p_1. \end{cases} \tag{4.5}$$

Here,

$$V_1 = \frac{a_0 + s - 1}{(a_0 a_1 - 1)(1 - s)},$$

$$V_2 = \frac{a_1(a_0 + s - 1)}{(a_0 a_1 - 1)}.$$

In Fig. 4.4, Eq. 4.5 is plotted with respect to several parameters. If t_1 and p_1 are now taken to be the respective proportions of T_1 and P_1 genes, depending

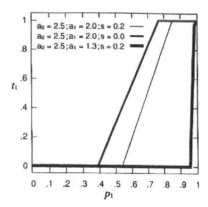

FIGURE 4.4: Kirkpatrick model and its equilibrium state (with permission of MIT Press [20]).

on the parameters and the initial conditions, the occurrence rate of T_1 takes an arbitrary value between 0 and 1. For both T_0 and T_1 to be preserved in the population, the following condition must be satisfied:

$$1 - a_0 < s < 1 - \frac{1}{a_1}, \qquad (4.6)$$

otherwise the population becomes stationary with respect to either T_0 or T_1.

The results indicate that an infinite number of equilibrium states exist for the occurrence rates of the genes. Hence, the equilibrium states lie on a curve rather than discrete points. Intuitively, natural selection should favor males carrying T_0; however, if the occurrence of P_1 is sufficiently high, T_1 males are preferred due to sexual selection, resulting in competition between natural and sexual selection.

In this model, changes in the occurrence rate of P depend only on the occurrence rate of T, which shows that once the population has entered an equilibrium state, it remains on this equilibrium path. However, this situation does not necessarily hold true when the occurrence rates of the genes are altered by external forces (such as genetic drift, migration, or sudden mutation).

4.1.3 Simulation using GAs

Collins and Jefferson simulated the Kirkpatrick model by using genetic algorithms [20], whereby each individual contained two chromosomes—one carrying gene P and the other carrying gene T. In females, only P was expressed, while in males only T was expressed, and each gene took one of two values (T_0 or T_1 for T and P_0 or P_1 for P). The population contained 131,072 individuals, with the same number of males and females at initiation. For each new generation, a fixed proportion of the T_1 males died before procreating,

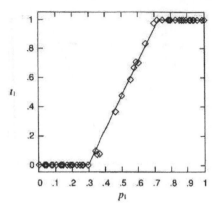

FIGURE 4.5: Evolutionary results after 500 generations (with permission of MIT Press [20]).

after which each female chose a mate from the remaining males, where mating consisted of gene recombination. Additionally, a sudden mutation could occur in all genes with a probability of 0.00001. This simulation weakened several assumptions of the Kirkpatrick model by assuming sudden mutations and a finite population size.

Figure 4.5 shows the results after 500 generations, where $a_0 = 2.0, a_1 = 3.0$, $s = 0.2$, $\mu = 0.00001$, and $N = 131,072$ are taken as the parameter values. The gene occurrence (of T_1 corresponding to P_1) in the final population is plotted over 51 iterations for various initial occurrences, and the results of Kirkpatrick's analysis are also included. The figure clearly shows that even with weaker assumptions the numerical results almost completely coincide with the analytical results.

Next, let us consider the evolutionary course until the population reaches equilibrium in the case where a new gene is inserted. Figure 4.6 shows the transitions for a population with an abundance of P_1 genes, but no T_1 genes. The initial generation thus had P_1 and T_1 proportions of $p_1 = 0.7$ and $t_1 = 0.0$, respectively, and the parameter values were $a_0 = 2.5, a_1 = 2.0, s = 0.2, \mu = 0.00001$, and $N = 131,072$. The rhomboids in the figure indicate the states for every 50 generations. Although 100 generations were required for a sufficient number of T_1 genes to emerge, after that point the population rapidly converged toward equilibrium.

This experiment is a remarkable demonstration of the power of female preferences and sexual selection. The females are less interested in the males whose survival is easier ($a_0 > a_1$), and since these T_0 males are not chosen as often, eventually the number of males carrying T_1, whose survival is more difficult, increases rapidly and dominates the population. Figure 4.7 shows that the average number of surviving T_0 males decreases as the number of T_1 males increases. This result is a clear demonstration of the "runaway effect" proposed

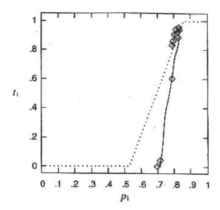

FIGURE 4.6: Evolutionary transition in the case where a new gene is inserted (with permission of MIT Press [20]).

by Fisher. Regardless of the fact that P and T are located in different chromosomes, a positive correlation is formed between P_1 and T_1. The increased preference for T_1 is also shown by the simulation, with the proportion of P_1 increasing from 0.7 to over 0.8.

Moreover, Collins and Jefferson conducted experiments with weakened assumptions. In one of these experiments a spatial restriction for choosing a mating partner was imposed (a more realistic model, since most animals do not mate with distant partners). In another experiment, diploidic organisms were provided with dominant and recessive genes. These two experiments are difficult to understand from a mathematical viewpoint, and in both experiments Collins and Jefferson found considerably different behavior compared with Kirkpatrick's original model. In the case where a spatial restriction was imposed on mating, T_1 spread slowly across the population; however, in the case of diploidic organisms, T_1 was unable to spread across the population unless the gene was prevalent from the beginning.

Apart from the research mentioned above, several other intriguing studies on sexual selection have been reported recently. For example, Werner examined the changing of characteristics in a population as a result of sexual selection, claiming that these changes are almost unlimited [124]. In addition, Werner and Todd showed that sexual selection in which detrimental features are preferred can result in the extinction of a species, and he also conducted simulations where evolution of the signal used by males to attract females was modeled through sexual selection [125]. Geoffrey Miller compared sexual selection to a venture company, and stated a theory emphasizing the possibility for progressive evolution that supports slow and conservative natural selection [85]. Furthermore, the application of sexual selection to engineering is also conceivable, for example, in interactive design systems, where certain

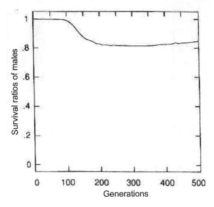

FIGURE 4.7: Survival ratios (male) vs. generations (with permission of MIT Press [20]).

types of runaway effect might prove to be productive. Such applications constitute an important topic for future research.

In this section, we introduced sexual selection together with several alternative theories as examples of simulations based on a GA. These arguments are not yet settled, and three of the theories presented (sexual selection, the handicap principle, and the parasite theory) cannot be considered in terms of being correct or incorrect. This matter is implied in the words of Darwin (see the quote at the beginning of this chapter).

4.2   Swarm-based simulation of sexual selection

Let us try to implement the simulation of gender selection based on the "Kirkpatrick" model in Swarm. In this model, the selection process for each generation is implemented as follows:

1. Each individual dies at a certain rate. Similarly, individuals die when they reach their life span.

2. At a rate corresponding to the gene frequency, the surviving females pair up with males and choose their traits (T gene).

3. Females mate with the male with chosen traits, and give birth to children. The child's genotype is determined by a genetic variation of the parents' genotype.

In this program, parameters to be specified are described in "`sexSelection.scm`" and are as follows:

- The initial group size: N

- The initial female male ratio (ratio of males to total population): male_femaleRatio

- The initial T genotype ratio (ratio of T_1 genotype to total population): tGeneratio

- The initial P genotype ratio (ratio of P_1 genotype to total population): pGeneRatio

- The probability of death of males of T_0 genotype: maleDeathProb

- The probability of extinction of males of T genotype: sRatio, males of T_1 genotype will survive with $(1-sRatio)$ times the probability as T_0 genotype males

- The probability of extinction of females: femaleDeathProb

- Mating probability a_0 by the Kirkpatrick model: a0Coeff

- Mating probability a_1 by the Kirkpatrick model: a1Coeff

- End generation of simulation: EndTime

- Lifetime of an individual: lifetime

- Children born by mating once: numChild

Executing the program displays the genotype frequency of each genotype and the number of individuals in each genotype (Fig. 4.8). Simulation stops when either males or females become extinct or the end generation is reached.

An individual's life and/or number of children can be specified by the parameters mentioned above. Furthermore, setting the extinction probability other than the T_0 genotype is also possible. Through this, the simulation of very reproductive individuals and individuals with longer life can be done.

Note that in the Kirkpatrick model, geographical conditions are not used, and basically random mating is done. Let's try to implement the gender selection process in a two-dimensional grid model in Swarm. In the simulation below, the rules of movement of an individual are as follows:

1. Female chooses one mating partner from among the males in sight (a square with side lengths of $2 \times$ visibility$+1$) depending on the frequency and a_0, a_1, and produces a child.

2. After that, the male and the female both move randomly.

The state of execution is shown in Fig. 4.9. The display in simulation and graphs is as follows:

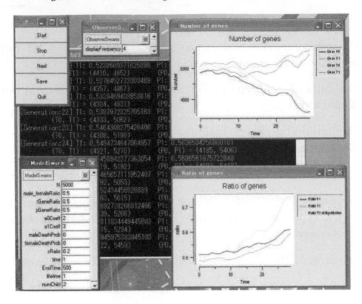

FIGURE 4.8: Swarm simulation based on Kirkpatrick's model.

Orange	`male_t0`	Male expressing the traits of T_0
Blue	`male_t1`	Male expressing the traits of T_1
Green	`female_p0`	Female expressing the traits of P_0
Yellow	`female_p1`	Female expressing the traits of P_1

4.3   Simulation of the prisoner's dilemma

4.3.1   The prisoner's dilemma

Two suspects (A and B) have been arrested by the police for a certain crime (Fig. 4.10). Although the suspects are accomplices and their guilt is strongly suspected, insufficient evidence exists to prove their guilt beyond doubt, and therefore the police separate the suspects and wait for a confession. A and B can adopt one of two strategies, namely, to confess (i.e., to defect, denoted by D below) or refuse to confess (i.e., to cooperate, denoted by C below), as shown in Table 4.1. The case where A does not confess is labeled A1, and the case where A confesses is labeled A2. Similarly for B, B1 denotes a refusal to confess and B2 denotes a confession. If neither suspect confesses, both of them would receive two-year sentences; however, if only one of them confesses, the one confessing would be sentenced to only one year due to extenuating circumstances, while the other suspect would receive a five-year

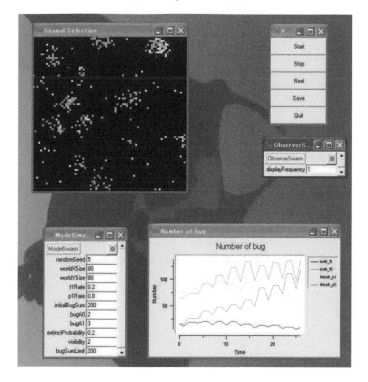

FIGURE 4.9 (See Color Insert): Sexual selection in a two-dimensional space.

sentence. Finally, if both suspects confess, both of them would receive three-year sentences.

Therefore, although the sentence is only two years for both suspects if neither of them confesses, the loss for one suspect is considerably greater if a confession is obtained from only their accomplice. The dilemma here is whether A and B should confess. In terms of pure strategy, it is straightforward to show that the pair (A2, B2) is an equilibrium point, and that no other equilibrium points exist. Here, the equilibrium point indicates that if either suspect adopts confession as their strategy, their accomplice is at a disadvantage if they do not adopt the same strategy. Conversely, if the suspects adopt the strategic pair (A1, B1), neither suspect confesses in the belief that their accomplice will not confess either, and the sentence for both of them is only two years. This sentence is obviously a more favorable outcome than the equilibrium point for both suspects, but since the suspects are separated from each other when taken into custody, they are unable to cooperate. This type of game is known as the prisoner's dilemma.

Situations such as those arising in the prisoner's dilemma are also often encountered in biology as well as in human society. For example, if an appro-

FIGURE 4.10: The prisoner's dilemma.

TABLE 4.1: Benefit and cost in the prisoner's dilemma.

(a) Benefit and cost from the viewpoint of A

	B_1 B: not confess (C)	B_2 B: confess (D)
A_1 A: not confess (C)	-2 Two years of imprisonment	-5 Five years of imprisonment
A_2 A: confess (D)	-1 One year of imprisonment	-3 Three years of imprisonment

(b) Benefit and cost from the viewpoint of B

	B_1 B: not confess (C)	B_2 B: confess (D)
A_1 A: not confess (C)	-2 Two years of imprisonment	-1 One year of imprisonment
A_2 A: confess (D)	-5 Five years of imprisonment	-3 Three years of imprisonment

priate incentive is present, the following situations can be modeled similarly to the prisoner's dilemma.

- Social interaction between animals in nature:
  - Social grooming in primates
  - Cleaner fish and the fish that they clean
  - Parasites and hosts

- Social interaction between humans:
  - Interaction between countries
  - Interaction between tribes

4.3.2 Iterated prisoner's dilemma

This section considers an extended version of the prisoner's dilemma known as the "iterated prisoner's dilemma" (IPD), in which the prisoner's dilemma is repeated a number of times with the same participants, and a final score is obtained as the sum of the scores in all iterations. Hereinafter, the choice made at each step is referred to as a move, or more precisely, a move indicates either cooperation (C: no confession) or defection (D: confession).

As with the prisoner's dilemma, situations similar to an IPD can regularly be observed in nature and in human society. A well-known example is regurgitation in vampire bats [25]. These small bats, living in Central and South America, feed on mammalian blood at night. However, their bloodsucking endeavors are not always successful, and at times they face starvation. Therefore, vampire bats that successfully fed will regurgitate part of their food and share it with other vampire bats that were unable to find food. Bats who receive food in this manner later return the favor. Wilkinson et al. observed 110 instances of regurgitation in vampire bats, of which 77 cases were from a mother to her children, and other genetic relations were typically involved in the remaining cases [120A]. Nevertheless, in several cases regurgitation was also witnessed between companions sharing the same den, with no genetic affinity. Modeling this behavior on the basis of the prisoner's dilemma, where cooperation is defined as regurgitation and defection is defined as the lack thereof, yields results similar to those in Table 4.2. A correlation between weight loss and the possibility of death due to starvation has also been found in bats. Therefore, the same amount of blood is of completely different value to a well-fed bat immediately after feeding and to a bat that has almost starved to death. In addition, it appears that individual bats can identify each other to a certain extent, and as a result they can conceivably determine how a certain companion has behaved in the past. Thus, it can be said that bats will donate blood to "old friends."

To set up the IPD, one player is denoted as P_1 and the other party is

TABLE 4.2: The strategy of vampire bats [25].

	Companion cooperates	Companion defects
I cooperate	**Reward: high** On the night of an unsuccessful hunt, I receive blood from you and avoid death by starvation. On the night of a successful hunt, I donate blood to you, which is an insignificant expenditure to me at that time.	**Compensation: zero** Although on a successful night I donate blood to you and save your life, you do not donate blood to me when my hunt is unsuccessful and I face death by starvation.
I defect	**Temptation: high** You save my life when my hunt is unsuccessful. However, on a successful night, I do not donate blood to you.	**Penalty: severe** I do not donate blood to you even when my hunt is successful, and death through starvation is a real threat to me on the night of an unsuccessful hunt.

TABLE 4.3: Codes and benefit chart in the IPD.

	The other party (P_2) does not confess $P_2 = C$	The other party (P_2) confesses $P_2 = D$
I (P_1) do not confess $P_1 = C$	Benefit: 3 Code: R	Benefit: 0 Code: S
I (P_1) confess $P_1 = D$	Benefit: 5 Code: T	Benefit: 1 Code: P

denoted as P_2. Table 4.1 is then used to create Table 4.3, describing the benefit for P_1. In this table, larger values indicate a higher benefit (which in the prisoner's dilemma is in the form of shorter imprisonment terms for the suspects).

Here, the following two inequalities define the conditions required for the emergence of the dilemma:

$$T > R > P > S, \tag{4.7}$$

$$R > \frac{T + S}{2}. \tag{4.8}$$

The first inequality is self-explanatory, while the implications of the second one will be explained later.

Strategies that can be considered for the IPD include

1. All-C: Cooperation (no confession) regardless of the move of P_2

2. All-D: Defection (confession) regardless of the move of P_2

3. Repeated cooperation and defection

4. RANDOM: a move is decided randomly for each iterate

The cooperative solution, which does not appear in a single game, evolves in the IPD. Cooperation becomes inevitable since defection entails subsequent revenge. An example of a well-known cooperation strategy is "tit for tat" (TFT):

1. The first move is performed at random.

2. All subsequent moves mirror P_2's move in the previous iteration.

In other words, this strategy follows the maxim "Do unto others as they have done unto you (an eye for an eye)." Thus, P_1 confesses if P_2 has confessed in their previous move, and vice versa. This strategy is recognized as being strong in comparison with other strategies (i.e., the total score is usually high). Variants of TFT include the following.

1. Tolerant TFT: Cooperate until P_2 defects in 33% of their moves.

2. TF2T: Defect only if P_2 has defected twice in a row.

3. Anti-TFT: The opposite strategy to TFT.

The benefit at each move and the total score are compared for All-C, All-D, TFT, and Anti-TFT strategies when pitted against each other in Table 4.4. The following can be inferred from the results presented in the table:

1. TFT is equivalent or superior to All-C.

2. TFT is generally superior to anti-TFT. However, anti-TFT is superior if P_2 adopts the All-C strategy.

3. All-C yields a lower score against Anti-TFT and All-D; however, it performs well against a TFT strategy.

4. All-D can outperform TFT.

In 1979, Axelrod extended an invitation to a number of game theorists and psychologists to submit their strategies for IPD [5]. Subsequently, a tournament was held using the 13 submitted strategies in addition to the RANDOM strategy. In this tournament, each strategy was set in turn against the others, including itself, in a round-robin pattern. To avoid stochastic effects, five matches were held for each strategic combination. Each match consisted of 200 iterations of the prisoner's dilemma in order to ensure a sufficiently long

TABLE 4.4: Comparison of the performance of the four strategies when pitted against each other.

Strategy of P_1	Strategy of P_2 (score/total score)			
	All-C	TFT	anti-TFT	All-D
All-C	3333/12	3333/12	0000/0	0000/0
TFT	3333/12	3333/12	0153/9	0111/3
Anti-TFT	5555/20	5103/9	1313/8	1000/1
All-D	5555/20	5111/8	1555/16	1111/4

interaction, as described below. Rather than considering the outcome of individual matches, the ultimate ranking was determined on the basis of the total score.

The winner of the contest was the TFT strategy proposed by Anatol Rapoport, a psychologist (and philosopher) at the University of Toronto. This program was the smallest of those submitted, consisting of only four lines of BASIC code. Upon analyzing the results, Axelrod derived the following two morals:

1. One should not deliberately defect from the other party (politeness).

2. Even if the other party defects, one should respond with defection only once and should not hold a grudge (tolerance).

Any strategy abiding by these morals is referred to as being "nice." Axelrod considered that a superior strategy could be derived based on these two morals. He held a second tournament, this time running a wide advertising campaign asking for entries through general computer magazines and publicizing the above two morals as a reference for the participants. Although this call for entries attracted a broad spectrum of participants from six countries, the TFT strategy proposed by Rapoport again emerged victorious, and Axelrod stated various other morals upon analyzing the results from both contests [5].

Looking at the scores obtained in the tournaments, nice strategies similar to TFT dominated the top half of the list. In contrast, the majority of the strategies in the bottom half of the list involved deliberate defection, while the lowest score was obtained by the RANDOM strategy. The question naturally arises as to why this set of circumstances occurs.

Let us consider the strength of the TFT strategy. First, we focus on the case where two nice strategies are competing. Since both strategies denounce deliberate defection, each of the 200 moves is one of cooperation. As a result, the two parties receive a total of $3 \times 200 = 600$ points. This situation is different when a nice strategy is competing against one that promotes defection. For example, consider competition between JOSS and TFT strategies. JOSS behaves almost identically to TFT; however, it defects on random occasions, corresponding to a player with a nice strategy occasionally being tempted to cheat. Thus, JOSS occasionally obtains five points by outperforming TFT (which supports cooperation and denounces defection), and although they are almost identical, JOSS appears to provide a greater benefit.

In the competition, TFT and JOSS cooperated for the first five moves.

```
TFT:   CCCCC
JOSS:  CCCCC
```

At this point, JOSS switched its tactics and defected (cheated).

```
TFT:   CCCCCC
JOSS:  CCCCCD
            ^
```

Although at the next move JOSS returned to its nice tactics, TFT took revenge by defecting.

```
TFT:   CCCCCCD
JOSS:  CCCCCDC
            ^
```

From here on, the moves of the two strategies alternated between defection and cooperation,

```
TFT:   CCCCCCDCDCD
JOSS:  CCCCCDCDCDC
            ^^^^^
```

and the respective scores at that juncture became

```
JOSS: 5 0 5 0 Ac
TFT:  0 5 0 5 Ac
```

giving an average of $\frac{0+5}{2} = 2.5$, which is less than the 3 points obtained if the two parties had continually cooperated (cf. the assumption in eq. 4.8).

Next, JOSS cheated again and defected on the 25th move,

```
TFT:   CCCCCCDCDCD...D
JOSS:  CCCCCDCDCDC...D
                   ^
```

resulting in reciprocal defection until the end of the match.

```
TFT:   CCCCCCDCDCD...DDDDDDD
JOSS:  CCCCCDCDCDC...DDDDDDD
                   ^^^^^^^
```

This chain of enmity yielded an average score of 1, and the match ended unchanged since neither side was able to forgive the other's defection and restore cooperation.

At the conclusion of the match, the total score was 241 to JOSS and 236 to TFT, lower than the 600 points that they could have obtained if they had cooperated throughout.

Hence, TFT has the following characteristics:

1. TFT never attacks the other party. The score obtained by TFT is the same or lower than that of the other party.

2. TFT does not hold a grudge. TFT seeks revenge only once, rather than punishing the other party repeatedly. In this way, TFT does not disregard the chance for reconciliation with the other party.

3. TFT does not require a victim, in other words, it does not exploit the other party. As a result, two players adopting the TFT strategy can enjoy a mutually beneficial relationship.

Conversely, although the All-D strategy can obtain the highest score of 5 points when competing against the All-C (victim) approach, the performance of using All-D is lower when competing against a different strategy.

An important assumption in the IPD is that the players do not know when the competition ends, otherwise it would allow for tactics based on "the ultimate defense measure" (always defecting on the last move). In addition, cooperative evolution requires a long competition since a relationship of trust, which is a prerequisite for cooperation, cannot be established in the short term. This idea can be explained using examples such as a student canteen (which adopts an honest business policy of expecting repeat clients in order to establish a long-standing relationship with students) and seaside noodle shops (which survive due to first-time customers, even with expensive and unappetizing food).

Next, we explain the concept of an evolutionarily stable strategy (ESS), whereby once a certain strategy becomes dominant in a population, invasion by another strategy is difficult. In the case of two competing strategies, A and B ($B \neq A$), by denoting the benefit associated with A as $E(A, B)$, the condition for an ESS is that either $E(A, A) > E(B, A)$ or both $E(A, A) = E(B, A)$ and $E(A, B) > E(B, B)$ are satisfied. Hence, All-C is obviously not an ESS. For example, if the All-D strategy is introduced into an All-C population, the entire All-C population would be destroyed by a single individual with the All-D strategy. Curiously, TFT is also not an example of an ESS.

Some disadvantages of TFT include

- In response to noise (when unintentional actions are made by mistake), TFT's moves begin to alternate between cooperation and defection.

- TFT cannot exploit the All-C strategy (since it usually cooperates).

TABLE 4.5: Comparison between Pavlov and TFT strategies.

preceding move P_1 P_2	Pavlov strategy	TFT strategy
C C	C	C
C D	D	D
D C	D	C
D D	C	D

To address these issues in TFT tactics, the so-called Pavlov strategy has been proposed, whereby P_1's current move is determined not only by the previous move of P_2, but also by P_1's preceding move (Table 4.5). For example, if an incorrect choice has been made (e.g., if P_1 cooperates or defects, expecting P_2 to cooperate, and instead P_2 defects), the previous choice is reversed in the current move. On the contrary, if P_2 cooperated while P_1 defected in the previous move, changing this choice in the current move is not necessary. The Pavlov strategy is considered to be robust with respect to noise. For example, in a competition where both players have adopted the Pavlov strategy, even if the cooperative relationship is broken due to noise and results in moves (C,D), a cooperative relationship is returned to via moves (D,D) followed by (C,C). Nevertheless, this does not imply that the Pavlov strategy is superior to the TFT strategy. For instance, although the TFT strategy consistently defects against the All-D strategy, the Pavlov strategy alternates between cooperation and defection, and its average score is lower than that for TFT.

IPD has been extended in various ways and has been actively used in research in the fields of AI, AL, and evolutionary economics [70]. Examples include studies in which strategic sexual selection is implemented by introducing the concept of gender, and studies where competition is extended to include three or more individuals, resulting in the emergence of cooperation and conspiracy.

Axelrod added norms to the prisoner's dilemma [6]. Here, defectors may be socially sanctioned by reduction of their score. Defecting players had a possibility to be detected by other players, and a strategy to decide whether to penalize a defector that was found (whether to reduce the score of the defector) was added to the strategy to play the prisoner's dilemma.

All players play one game with all other players in each generation. Every time a player defects, its defection may be detected by other players in the population. The detected defector is penalized according to the detector's tendency to retaliate, which is given as a probability.

The evolution process is applied after all battles finish. The strategy of children may change by mutation, and children may have boldness and a strong tendency to retaliate that are different from their parents. Setting no norms would result in the entire population becoming defectors.

In addition, Axelrod added metanorms, which is a norm to penalize detectors that do not penalize defectors, because collaborators will not always evolve when only norms exist. When players that penalize other players who forgive defectors were introduced, there was a tendency for former players to evolve to penalize defectors and for defectors that were penalized to become cooperative.

4.3.3   IPD using GAs

This section describes an experiment based on research conducted by Cohen et al. [19], where an IPD strategy evolves by using a GA. As before, the first player is denoted as P_1 and the other party is denoted as P_2, and the benefit chart in Table 4.3 is used to score the moves.

In this experiment, errors (i.e., when an unintentional move is made) and noise (where the move communicated to the opponent is different from the actual move) are introduced, and the emergence of cooperation in the case of incomplete information is examined. There are 256 players in the IPD, arranged on a 16×16 lattice, and the competition between players consists of four iterations. A single experiment consists of 2500 periods, where a single period comprises the following steps:

- Competitions between all players

- Learning and evolution of strategies in accordance with an appropriate process

The basic elements of the experiment are as follows.

1. Strategy space—an expression of the strategy adopted by a player, which is divided into the following two types.

   (a) Binary: pure strategy

   (b) Continuous: mixed strategy (stochastic strategy)

2. Interaction process—the set of opponents competing with a player is referred to as the neighborhood of that player. The neighborhood can be chosen in six ways.

   (a) 2DK (2 Dimensions, Keeping): the neighbors above, left, below, and right (NEWS: north, east, west, and south) of each player on the lattice are not changed, and the neighborhood is symmetrical. In other words, if B is in the neighborhood of A, then A is necessarily in the neighborhood of B.

   (b) FRNE (Fixed Random Network, Equal): for each agent, there is a neighborhood containing a fixed number of agents (four). These agents are selected at random from the entire population and remain fixed until the end. The neighborhood is symmetrical.

(c) FRN (Fixed Random Network): almost the same as FRNE, although with a different number of agents in the neighborhood of each agent. The neighborhood is unsymmetrical. The average number of agents is set to 8.

(d) Tag: a real number between 0 and 1 (a tag or sociability label) is assigned to each agent, introducing a bias whereby agents with similar tags can more easily compete with each other.

(e) 2DS (2 Dimensions, 4 NEWS): the same as 2DK, except that the agents change their location at each period; thus NEWS is set anew.

(f) RWR (Random-With-Replacement): a new neighborhood can be chosen at each period. The average number of agents in the neighborhood is set to 8, and the neighborhood is unsymmetrical.

3. Adaptation Process: the evolution of a strategy can follow any of the following three paths.

(a) Imitation: searching the neighborhood for the agent with the highest average score and copying its strategy if this score is larger than one's own score.

(b) BMGA (Best-Met-GA): almost the same as Imitation; however, an error is generated in the process of copying. The error can be any of the following.

    i. Error in comparing the performance: the comparison includes an error of 10%, and the lower estimate is adopted.

    ii. Miscopy (sudden mutation): the mutation rate for genes in the mixed strategy (Continuous) is set to 0.1, with the addition of Gaussian noise (mean, 0; standard deviation, 0.4). In addition, the rate of sudden mutation per gene in the pure strategy (Binary), in which probabilities p and q are flipped (i.e., 0 and 1 are swapped), is set to 0.0399. When p mutates, probability i is also changed to the new value of p.

(c) 1FGA (1-Fixed-GA): BMGA is applied to agents selected at random from the entire population.

The strategy space is expressed through the triplet (i, p, q), where i is the probability of making move C at the initial step, p is the probability of making move C when the move of the opponent at the previous step was C, and q is the probability of making move C when the move of the opponent at the previous step was D.

By using this representation, various strategies can be expressed as follows:

$$
\begin{array}{ll}
\text{All-C} & i = p = 1,\ q = 1 \\
\text{All-D} & i = p = 0,\ q = 0 \\
\text{TFT} & i = p = 1,\ q = 0 \\
\text{Anti-TFT} & i = p = 0,\ q = 1
\end{array}
$$

Here, the value of each of i, p, and q is either 0 or 1. This form of representation of the strategy space is known as a binary (or pure) strategy.

Conversely, the values of p and q can also be real numbers between 0 and 1, and this representation is known as a continuous strategy. In particular, since 256 players took part in the experiment conducted by Cohen, the initial population was selected from among the following combinations:

$$p \;=\; \frac{1}{32}, \frac{3}{32}, \cdots, \frac{31}{32}, \tag{4.9}$$

$$q \;=\; \frac{1}{32}, \frac{3}{32}, \cdots, \frac{31}{32}. \tag{4.10}$$

Specifically, each of the 16×16 combinations is set as a single agent (player) with values p, q. Moreover, the value of i is set to that of p.

The performance (the average score of all players in each competition) is calculated as the average value over 30 runs. Note that the occurrence of defection is high if the average value of the population performance is close to 1.0, while cooperation dominates if the value is close to 3.0. Nevertheless, the value does not necessarily reach 3.0 even if the entire population adopts the TFT strategy due to (1) sensor noise in the comparison and (2) copying errors in the adaptation process.

The score of the population can be obtained from the following measures:

- Av1000: the average score of the last 1000 periods

- NumCC: the number of periods where the average score for the entire population was above 2.3

- GtoH: out of 30 iterations, the number of iterations in which the average score for the entire population was above 2.3

- FracH: the proportion of periods with a score exceeding 2.3 after the average score for the entire population has reached 2.3

- FracL: the proportion of periods with a score below 1.7 after the average score for the entire population has reached 2.3

- Rstd: the averaged score distribution for each period during iteration

Here, an average score of 2.3 is regarded as the threshold above which the population is cooperative, and a score below 1.7 indicates a population is dominated by defection.

The results for all interaction and adaptation processes, in addition to all combinations of representations in the experiments ($2 \times 6 \times 3 = 36$ combinations), are shown in Tables 4.6 and 4.7.

In conducting the experiments, an oscillatory behavior at the cooperation level was observed for all combinations, namely, the average score oscillated between high (above 2.3) and low (below 1.7) cooperation. This oscillation can be explained from repetition of the following steps.

1. If a TFT player happens to be surrounded by TFT players, defecting players are excluded.

2. TFT players soon establish a relationship of trust with other TFT players, and in the course of time the society as a whole becomes considerably more cooperative.

3. Defecting players begin to exploit TFT players.

4. Defecting players take a course to self-destruction after exterminating the exploited players.

5. Return to 1.

Two important factors exist in the evolution of cooperation between players. The first factor is preservation of the neighborhood in the interaction process, where the neighborhood is not limited to two-dimensional lattices. A cooperative relationship similarly emerges in fixed networks, such as FRN. In addition, agents that stochastically select their neighborhoods by using tags are more cooperative than agents with random neighborhoods.

The second factor is the process of generating and maintaining the versatility of the strategy. For example, cooperation is not common in adaptation processes that are similar to Imitation. Since errors do not exist in Imitation, the versatility of the strategy is soon lost, which often results in an uncooperative population. However, on rare occasions, extremely cooperative populations arise even in Imitation, and Imitation was the strategy producing the highest scoring population.

4.3.4 IPD simulation by Swarm

This section describes the Swarm simulation of Cohen's iterated prisoner's dilemma. Figure 4.11 shows the state of execution screen. The player is placed in the two-dimensional grid of `worldSize`×`worldSize`. `worldSize` is set to 16 (inside the file `GridPD.smc`).

According to Cohen's definition, this IPD's basic elements are as follows:

1. Strategy Space: Binary, actually of only four types: All-C, TFT, Anti-TFT, All-D.

2. Iteration Process: 2DK, vertical-horizontal agent is set as the neighborhood. Sending the method "`setNeighbourhood`" for each player "`aPlayer`" sets this. If we look at the definition of the method "`setNeighborhood`" inside "`modelSwarm.java`"

```
setNeighborhood$atDX$DY(aPlayer,-1,0);
setNeighborhood$atDX$DY(aPlayer,1,0);
setNeighborhood$atDX$DY(aPlayer,0,-1);
setNeighborhood$atDX$DY(aPlayer,0,1);
```

TABLE 4.6: IPD benefit chart and codes (binary representation).

Interaction process	Adaptation process	Av1000	NumCC	GtoH	FracH	FracL	Rstd
2DK	IFGA	2.560 (.013)	30	19	.914	.001	0.173 (.017)
	BMGAS	2.552 (.006)	30	10	.970	.000	0.122 (.006)
	ImitS	3.000 (.000)	30	7	1.00	.000	0.000 (.000)
FRNE	IFGA	2.558 (.015)	30	21	.915	.000	0.171 (.012)
	BMGAS	2.564 (.007)	30	9	.968	.000	0.127 (.007)
	ImitS	2.991 (.022)	30	6	1.00	.000	0.000 (.000)
FRN	IFGA	2.691 (.008)	30	22	.990	.000	0.120 (.010)
	BMGAS	2.629 (.010)	30	14	.913	.006	0.229 (.016)
	ImitS	1.869 (1.01)	13	7	1.00	.000	0.000 (.000)
Tag	IFGA	2.652 (.010)	30	15	.975	.000	0.132 (.010)
	BGMAS	1.449 (.186)	30	255	.191	.763	0.554 (.170)
	ImitS	1.133 (.507)	2	6	1.00	.000	0.000 (.000)
2DS	IFGA	2.522 (.024)	30	26	.867	.000	0.197 (.020)
	BMGAS	2.053 (.128)	30	95	.443	.280	0.532 (.064)
	ImitS	1.000 (.000)	0	-	-	-	0.000 (.000)
RWR	IFGA	2.685 (.0009)	30	54	.985	.000	0.127 (.013)
	BMGAS	1.175 (.099)	13	972	.191	.763	0.109 (.182)
	ImitS	1.000 (.000)	0	-	-	-	0.000 (.000)

TABLE 4.7: IPD benefit chart and codes (continuous representation).

Interaction process	Adaptation process	Av1000	NumCC	GtoH	FracH	FracL	Rstd
2DK	1FGA	2.025 (.069)	30	127	.213	.172	0.313 (.045)
	BMGAS	2.554 (.009)	30	26	.998	.000	0.075 (.006)
	ImitS	2.213 (.352)	15	27	.868	.000	0.016 (.003)
FRNE	1FGA	2.035 (.089)	30	122	.226	.161	0.292 (.049)
	BMGAS	2.572 (.007)	30	26	.996	.000	0.077 (.007)
	ImitS	2.176 (.507)	14	26	.901	.000	0.017 (.014)
FRN	1FGA	1.884 (.120)	30	162	.182	.311	0.379 (.044)
	BMGAS	2.476 (.026)	30	40	.939	.003	0.124 (.062)
	ImitS	1.362 (.434)	3	12	1.00	.000	0.009 (.004)
Tag	1FGA	2.198 (.057)	30	80	.380	.064	0.248 (.058)
	BGMAS	1.613 (.277)	30	331	.117	.499	0.469 (.093)
	ImitS	1.573 (.187)	0	-	-	-	0.012 (.004)
2DS	1FGA	1.484 (.086)	30	695	.040	.764	0.273 (.080)
	BMGAS	1.089 (.003)	1	978	.001	.977	0.024 (.009)
	ImitS	1.096 (.013)	0	-	-	-	0.006 (.000)
RWR	1FGA	1.502 (.109)	30	612	.055	.729	0.332 (.064)
	BMGAS	1.098 (.036)	9	1384	.021	.883	0.063 (.097)
	ImitS	1.104 (.032)	0	-	-	-	0.006 (.001)

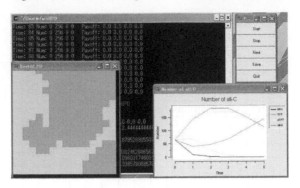

FIGURE 4.11: IPD simulation.

it is as above. We can define another iteration process by rewriting this part.

3. Adaptive Process: Imt. (Imitation), in other words, looks for the agent with the largest average gain among the agents, and copies it if it has larger average gain than itself. Sending the method `adaptType` for each player "`aPlayer`" sets this.

Keeps/holds the four types of strategies (All-C, TFT, Anti-TFT, All-D) agents in `num[0]`, `num[1]`, `num[2]`, and `num[3]`, respectively. This program keeps cooperation (C) and deception (D) as 0. Let us look at the declaration inside "`Player.java`."

```
static final int iParam[] = {1, 1, 0, 0};
static final int pParam[] = {1, 1, 0, 0};
static final int qParam[] = {1, 0, 1, 0};
```

Here, `iParam` works first, `pParam` works next once before the time of C, `qParam` works next once before the time of D. Again, 0 to 3 of the array corresponds to the four types of strategies (All-C, TFT, Anti-TFT, All-D). For example, `iParam[0]`, `pParam[0]`, and `qParam[0]` are the strategies of All-C. Recall that each agent's numbers are kept in `num[0]`, `num[1]`, `num[2]`, and `num[3]`. `num[0]`, `num[1]`, `num[2]`, and `num[3]` values (1/4 each by default) are set at the initialization part of the model (`buildObjects` of `ModelSwarm.java`). After this, the list of players is initialized and placed in the two-dimensional grid. Note that the following method swaps the order of the list randomly (if this is not done, they will be placed in "`world`" with the following order: All-C, TFT, Anti-TFT, All-D).

```
shuffle(playerList);
```

The four values `num[0]`, `num[1]`, `num[2]`, and `num[3]` are displayed on a line graph (`EZGraph`) for each period. Their colors, as described in the definition, are red, blue, orange, and black, respectively. This display is done in the following part by the `buildObjects` method of the `ObserverSwarm`:

```
numGraph.createSequence$withFeedFrom$andSelector(
    "all-C",modelSwarm,new Selector(
    modelSwarm.getClass(),"getNum0", false));
numGraph.createSequence$withFeedFrom$andSelector(
    "TFT",modelSwarm,new Selector(
    modelSwarm.getClass(),"getNum1", false));
numGraph.createSequence$withFeedFrom$andSelector(
    "aTFT",modelSwarm,new Selector(
    modelSwarm.getClass(),"getNum2", false));
numGraph.createSequence$withFeedFrom$andSelector(
    "all-D",modelSwarm,new Selector(
    modelSwarm.getClass(),"getNum3", false));
```

Here, "setNum0," etc., are the methods that return the number of array elements (`num`).

Each player keeps its strategy type in "`type`" (integer value from 0 to 3). This corresponds to the four types of strategies above (All-C, TFT, Anti-TFT, All-D). By sending the "`step`" method to the player's object, a hand is obtained, and is substituted in the internal variable "`newAction`" inside `Player.java`, the definition of "`step`" is as follows:

```
public void step(int t){
    numPlays++;
    if(t==0) // is it the first hand?
        newAction=iParam[type];
    else
        if(memory==1) // is C the hand of the previous opponent?
            newAction=pParam[type];
        else
            newAction=qParam[type];
}
```

In the "`memory`" variable, the previous opponent's hand is kept. Calling the "`remember`" method after each play sets this variable.

The first play is defined in "`runTournament`" of `ModelSwarm.java`. Inside this, for the following method, each player `player` executes the play for each neighborhood (`neigh`).

```
runTournament$against(player,neigh);
```

The play part is defined as follows in `Tournament.java`:

```
public Object run(){
    int t;
```

```
    numIter = 4;   // let the play be of four rounds
    for (t=0; t<numIter; t++) {
        updateMemories(); // stores the hand of previous opponent
        player1.step(t);  // obtains the hand of player 1
        player2.step(t);  // obtains the hand of player 2
        distrPayoffs();   // calculates the gain from play result
    }
    return this;
}
```

4.3.5   IPD as spatial games

The iterated prisoner's dilemma (IPD) is being extensively researched in various fields.

When the prisoner's dilemma is configured as a spatial game, it results in interesting patterns such as the "big bang" or the "kaleidoscope" [93].

The rules of the spatial game are as follows. Each player occupies a point on a two-dimensional lattice, and battles with all adjacent players. The scores from these battles are added to obtain the gain. Each player on a lattice point changes its strategy to that of its opponent with the highest gain.

The players are displayed in the following colors:

- Blue: C in last generation, C in current generation

- Red: D in last generation, D in current generation

- Green: D in last generation, C in current generation

- Yellow: C in last generation, D in current generation

Figure 4.12 shows a kaleidoscope obtained from evolution with Swarm. The parameter probe in ModelSwarm has the following meanings:

- isBigbang: select big bang if 1

- isKaleidoscope: select kaleidoscope if 1

The kaleidoscope appears from a single defector that enters a square of cooperators of fixed size. Rows of symmetric patterns that continuously change appear for a surprisingly long period. The number of possible configurations is finite; therefore the kaleidoscope reaches a stable pattern or cycle after a long period of time.

Huberman and Glance expanded this model by adding asynchronous elements in this spatial model [53]. They chose the procedure of the game such that selection is carried out after a number of battles between nearby groups, and again after battles between different groups. As a result, such a simple change typically results in the entire plane becoming defectors.

4.4   Evolving artificial creatures and artificial life

4.4.1   What is artificial life?

Artificial Life (AL) aims to solve fundamental problems of biology on a computer. The first Artificial Life Symposium was held at the Santa Fe Institute in the fall of 1987 [78]. Christopher Langton, the organizer of this symposium, defined the credo of artificial life researchers as "those who pursue ghosts in a machine, or the nature of what comes out of, but is independent of materials."

AL is difficult to define, but the following outlines the basic features of AL in computers [78, p. 3]:

- They consist of populations of simple programs or specifications.

- There is no single program that directs all of the other programs.

- Each (microscopic) program details the way in which a simple entity reacts to local situations in its environment, including encounters with other entities.

- There are *no* rules in the system that dictate global (macroscopic) behavior.

- Any behavior at levels higher than the individual programs is therefore emergent.

The term "emergent property" was originally used in biology, and a good example is the social behavior of ants. Individual ants perform only simple mechanical actions; however, the ant colony as a whole performs highly intellectual collective behavior based on the distribution patterns of food and enemies to increase the survival rate of the entire group. As a result, a caste

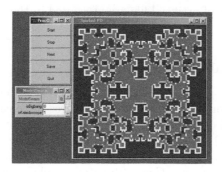

FIGURE 4.12: Evolutionary kaleidoscope.

system of specialized individuals that do different work is formed, and social division of labor and cooperative phenomena arise. This is an example of an emergent property because no rule (program) exists that governs the action of the entire colony, but the collective action of simple programs (individual ants) results in intellectual action of the entire group.

Artificial life aims to study "behavior like that in organisms" and uses synthetic research approaches rather than analytic methods to investigate life phenomena. The significance of discovering a mechanism that results in the emergence of behavior in organisms is two-fold:

- Scientific significance: Deeper understanding of existing life, fundamental properties of life in particular.

- Engineering significance: Formation of an artificial product that can adapt to its surroundings.

Artificial life researchers aim to model complex phenomena, such as life, on a computer. Modeling is expected to accurately predict and control the behavior of organisms and be utilized in engineering applications.

Artificial life aims to expand conventional biology regarding biological life (Blife) and establish a new branch of evolutionary biology that is verifiable through simulation or other methods. The difference between Alife and Blife can be summarized as

Alife: Life as it could be

Blife: Life as we know it

Kauffman recently expanded the methodologies used in artificial life and proposed the concept of "general biology" and the hypothesis that life has been building a search space (fitness landscape) such that the search process will succeed [62]. The validity of this hypothesis is not known, but an interesting task in the future would be to re-investigate evolutionary computational methods from this viewpoint.

Some examples of artificial life are presented in the following sections and chapters.

4.4.2   Artificial life of Karl Sims

Karl Sims did pioneering work on the "artificial evolution" of artificial life [106]. Virtual organisms are created in three-dimensional space on a computer. The virtual organisms have genes represented by directed graphs, and their shape (body) consists of multiple blocks as the expression type. The organisms have a number of sensors, and output some action based on the input (output to effectors). Genes (directed graphs) consist of intranode connection information of nodes that represent component blocks (Fig. 4.13). The

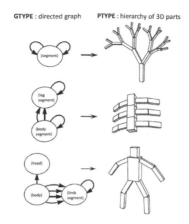

FIGURE 4.13: Designed examples of genotype graphs and corresponding creature morphologies [106].

graphs are neural networks that control multiple blocks and act as stimulus-and-reaction systems that generate action based on sensor input.

Physical simulations were used to calculate the motion of bodies, numerical integration, detection of collisions, and collision reaction from friction to achieve evolution of virtual life. This follows the standard methodology of virtual reality, and allowed accurate fitness calculations in virtual space. Genetic operators in evolution included grafting in addition to mutation and crossover of networks that are used in standard GP.

Virtual organisms with various shapes and actions appeared through evolutionary processes in response to problems such as "walking," "swimming" (Fig. 4.14), and "jumping." The evolutionary process was reminiscent of the Cambrian Explosion. Furthermore, co-evolution of virtual organisms was observed (Fig. 4.15) where generated organisms fought against each other for common resources, and only the winner survived and reproduced.

4.4.3 Evolutionary morphology for real modular robots

The author has extended Sims's approach to evolving real-world robots and successfully applied EC to the morphology for cubic creatures [117, 118]. We proposed a new approach based on a GA and CA (cellular automata; see Chapter 7) to automatically building patterns of block-type elements. We observed the emergence of the effective patterns in both virtual and real worlds, some of which seemed to be surprisingly counterintuitive.

One experiment was made to attempt the evolutionary morphology of the working leg parts instead of wheels, without using sensors. Fitness was computed by the moving distance on the plane multiplied by the stable movement time.

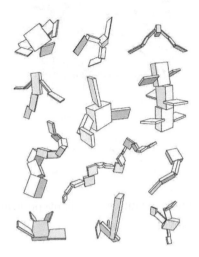

FIGURE 4.14: Creatures evolved for swimming [106].

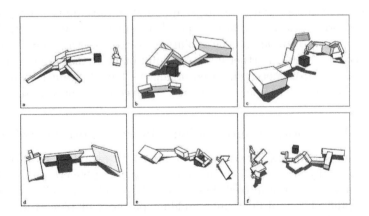

FIGURE 4.15: Evolved competing creatures (with permission of MIT Press [108]).

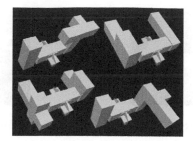

FIGURE 4.16: Various individuals.

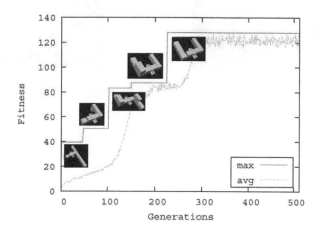

FIGURE 4.17: Best individuals with generations.

Figures 4.16 and 4.17 show some of the individuals that emerged during the evolution, and Fig. 4.18 shows the second optimal individual resulting from the evolution. The optimal individual uses the leg part as a support during rotation, and the central rollers handle the rotating movement. We performed the same experiment with the actual machines and observed the behavior in the real world.

It may be easy to imagine individuals that adopt a controller of constant rotation, maximizing the diameter of the leg parts; however, an individual as in Fig. 4.18 was actually obtained. The moving distance was maximized by rotation using the rollers attached to the body, which was realized by spreading the leg part in one direction. In reality, the individuals that maximize the diameter, such as the one with the cross-shaped feet (Figure 4.19), cannot take advantage of friction as a driving force, thereby spinning around and going nowhere. We performed the same experiment with the actual machines and the same tendency was also seen.

FIGURE 4.18: The suboptimal individual and the movement.

FIGURE 4.19: Cross-shaped robot.　　**FIGURE 4.20**: Wheel robot.

From the viewpoint of the practical movement method of modular robots, they can be classified into those that transform like ameba to change a whole location and those that use leg parts or wheels (Figure 4.20). Although this result is an option due to a higher velocity than wheels, it requires sufficient strength and a function to shift directions.

Another experiment was attempted to ascend stairs via morphogenesis using GP.

Figure 4.21 shows the "good" individuals that emerged in each generation. The light-colored blocks in the center are normal blocks, and the dark-colored blocks at the edge are motor blocks.

Figure 4.21(a) shows an individual that emerged in the third generation. Because only one motor emerged, which made proceeding straight ahead difficult, it was not able to achieve a meaningful task. Figure 4.21(b) shows an individual that emerged in the eighth generation, featuring five vertically aligned blocks.

The robot supported itself using its long vertical part and got on the gap between the stairs, but the tilt could not be supported. Afterward, the robot deviated from the right orbit. The individual that emerged in the 14th generation, shown in Fig. 4.21(c), ascended the stairs, supported by the five central blocks as well. Starting with the eighth generation, there were two more blocks at the front. The wider shape could avoid deviation from the orbit caused by the tilting at the time of climbing over the gap in the stairs.

Figure 4.21(d) shows the best individual during the simulation, which emerged in the 16th generation. Putting the three blocks on the right-side front of the robot upon the next stair, the robot could start climbing the staircase. Compared with Fig. 4.21(c), there were fewer front blocks, thereby

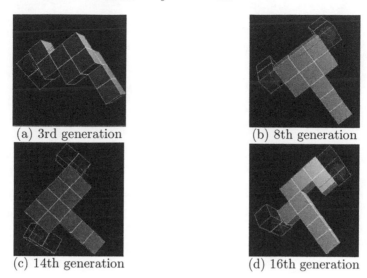

(a) 3rd generation (b) 8th generation

(c) 14th generation (d) 16th generation

FIGURE 4.21: The best evolved individuals.

eliminating the front block obstacles and allowing the robot to ascend over the gap in the stairs and shorten the time to realize the task.

Chapter 5

Ant Colony–Based Simulation

> We enter my study, candle in hand. One of the windows had
> been left open, and what we see is unforgettable. With a soft
> flick-flack the great Moths fly around the bell-jar, alight, set
> off again, come back, fly up to the ceiling and down. They
> rush at the candle, putting it out with a stroke of their wings;
> they descend on our shoulders, clinging to our clothes, grazing
> our faces [37, pp. 168–169].

5.1 Collective behaviors of ants

Ants march in a long line. There is food at one end, a nest at the other. This
is a familiar scene in gardens and on roads, but the sophisticated distributed
control by these small insects was recognized by humans only a few decades
ago.

Ants established their life in groups, or colonies, more than a hundred
million years before humans appeared on Earth. They formed a society that
handles complex tasks such as food collection, nest building, and division of
labor through primitive methods of communication. As a result, ants have
a high level of fitness among species, and can adapt to harsh environments.
New ideas including routing, agents, and distributed control in robotics have
developed based on simple models of ant behavior. Applications of the ant
behavior model have been used in many papers, and are becoming a field of
research rather than a fad.

Marching is a cooperative ant behavior that can be explained by the
pheromone trail model (Fig. 5.1). Cooperative behavior is frequently seen
in ant colonies, and has attracted the interest of entomologists and behavioral
scientists. Pheromones are volatile chemicals synthesized within the insect,
and are used to communicate with other insects of the same species. Examples
are sex pheromones that attract the opposite sex, alarm pheromones that
alert group members, and trail pheromones that are used in ant marches.
Pheromones are discussed in a chapter of the well-known *Souvenirs Entomologiques* by Jean-Henri Fabre (see the quote at the beginning of this chapter).

However, recent research indicates that pheromones are effective within a

FIGURE 5.1: Ant trail.

distance of only about 1 m from the female. Therefore, it is still not known if males are attracted only because of the pheromones.

Many species of ants leave a trail of pheromones when carrying food to the nest. Ants follow the trails left by other ants when searching for food. Pheromones are volatile matter that is secreted while returning from the food source to the nest. The experiments shown in Fig. 5.2 by Deneubourg and Goss using Argentine ants linked this behavior to the search for the shortest path [43]. They connected bridge-shaped paths (two connected paths) between the nest and the food source, and counted the number of ants that used each path. This seems like a simple problem, but because ants are almost blind they have difficulty recognizing junctions, and cannot use complex methods to communicate the position of the food. Furthermore, all the ants must take the shorter path to increase the efficiency of the group. Ants handle this task by using pheromones to guide the other ants.

Figure 5.3 shows the ratio of ants that used the shorter path [43]. Almost every ant used the shorter path as time passed. Many of the ants return to the shorter path, secreting additional pheromones; therefore, the ants that followed also take the shorter path. This model can be applied to the search for the shortest path, and is used to solve the traveling salesman problem (TSP) and routing of networks. There are many unknown factors about the pheromones of actual ants; however, the volatility of pheromones can be utilized to build a model that maintains the shortest path while adapting to rapidly changing traffic. The path with a greater accumulation of pheromones is chosen at junctions, but random factors are inserted to avoid inflexible solutions in a dynamic environment.

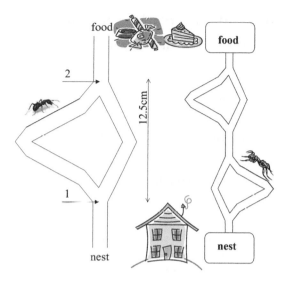

FIGURE 5.2: Bridge-shaped paths (adapted from Fig. 1a in [43], with permission of Springer-Verlag GmbH).

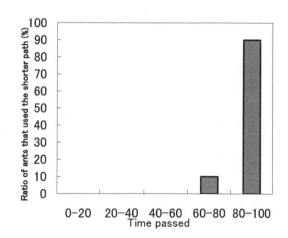

FIGURE 5.3: The ratio of ants that used the shorter path (adapted from Fig. 1a in [43], with permission of Springer-Verlag GmbH).

5.2   Swarm simulation of the pheromone trails of ants

An easy model can describe the actions of ants as follows:

- In the case of nothing, a random search is done.

- If the food is found, it takes it back to the hive. A homing ant knows the position of the hive, and returns almost straight back.

- Ants that take the food back to the hive drop their pheromone. Pheromones are volatile.

- Ants not having the food have the habit of being attracted to the pheromone.

Figure 5.4 is the execution state in Swarm. Here, the hives are placed in the center, and there are three (lower right, upper left, lower left) food sources. Figure 5.4(a) is the first random search phase. In (b), the closer lower right and lower left food is found, and the pheromone trail is formed. The upper left is in the middle of the formation. In (c), pheromone trails are formed for all three sources, which makes the transport more efficient. The lower right source is almost exhaustively picked. In (d), the lower right food source finishes, and the pheromone trail is already dissipated. As a result, a vigorous transportation for the two sources on the left is being done. After this, all the sources finish, and the ants return to random search again. The parameters in the simulation are shown in Table 5.1.

In this program, at the time of stopping, food locations and various parameters can be changed dynamically. For this purpose, the **probe** method is used, as described in Section 3.2.7. Specific operations are as follows:

- To change evaporation and diffusion coefficients: Enter the variable and press "enter." Then click "`initializeEvaporationAndDiffusionRate`."

- To change the bugs' parameters or colony size: Enter the variable and press "enter." Then click "`initializeBugAndColonySize`."

- To change foods' positions: Enter x, y coordinates and press "enter." Then click "`initializeFood`."

- To add a food source: Enter x, y coordinates and radius and press "enter." Then click "`initializeFood`."

- To delete a food source: Enter the minus value in the food's radius and press "enter." Then click "`initializeFood`."

Let us try and check how the ants' behavior changes when the food location is changed. Especially, how robust is the search using pheromones under disturbances?

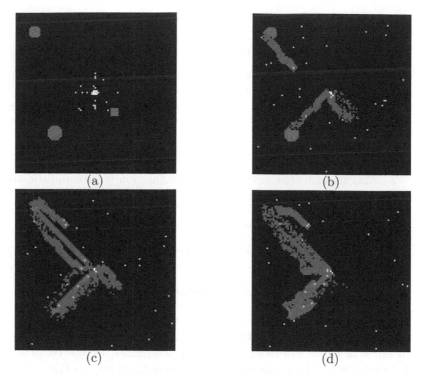

(a) (b)

(c) (d)

FIGURE 5.4 (See Color Insert): Pheromone trails of ants.

TABLE 5.1: Parameters of the pheromone trails of ants.

Parameter	Meaning	Range
`amountOfReleasingPheromone`	Amount of pheromone dropped.	0∼
`evaporationRate`	Ratio of pheromone that evaporates to the amount dropped on the ground.	0 1
`diffusionRate`	Proportion of the evaporated pheromone that diffuses.	0∼0.2
`awayFromColonyRate`	The proportion of priority given to the pheromone away from the hive. Ignores the direction of hive when 1.	1∼
`turnRate`	The proportion of not going straight when searching for food. Becomes a random walk when 1.	0∼1

5.3   Ant colony optimization (ACO)

Optimization algorithms based on the collective behavior of ants are called ant colony optimization (ACO) [31].

ACO using a pheromone trail model for the TSP uses the following algorithm to optimize the travel path.

1. Ants are placed randomly in each city.

2. Ants move to the next city. The destination is probabilistically determined based on the information on pheromones and given conditions.

3. Repeat until all cities are visited.

4. Ants that make one full cycle secrete pheromones on the route according to the length of the route.

5. Return to 1 if a satisfactory solution has not been obtained.

The ant colony optimization (ACO) algorithm can be outlined as follows. Take η_{ij} as the distance between cities i and j. The probability $p_{ij}^k(t)$ that an ant k in city i will move to city j is determined by the reciprocal of the distance $1/\eta_{ij}$ and the amount of pheromone $\tau_{ij}(t)$ (eq. (5.1)):

$$p_{ij}^k(t) = \frac{\tau_{ij}(t) \times \eta_{ij}^\alpha}{\sum_{h \in J_i^k} \tau_{ih}(t) \times \eta_{ih}^\alpha}. \tag{5.1}$$

Here, J_i^k is the set of all cities that the ant k in city i can move to (has not visited). The condition that ants are more likely to select a route with more pheromone reflects the positive feedback from past searches as well as

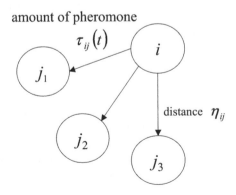

FIGURE 5.5: Path selection rules of ants.

TABLE 5.2: Comparison between ACO and metaheuristics.

TSP	ACO	GA	EP	SA	Optimal
Oliver 30	420	421	420	424	420
	[830]	[3200]	[40,000]	[24,617]	
Eil 50	425	428	426	443	425
	[1,830]	[25,000]	[100,000]	[68,512]	
Eil 75	535	545	542	580	535
	[3,480]	[80,000]	[325,000]	[173,250]	
KroA 100	21,282	21,761	N/A	N/A	21,282
	[4,820]	[103,00]	[N/A]	[N/A]	

a heuristic for searching for a shorter path. The ACO can thereby include an appropriate amount of knowledge unique to the problem.

The pheromone table is updated by the following equations:

$$Q(k) \quad = \quad \text{the reciprocal of the path that ant } k \text{ found} \qquad (5.2)$$

$$\Delta\tau_{ij}(t) \quad = \quad \sum_{k \in A_{ij}} Q(k) \qquad (5.3)$$

$$\tau_{ij}(t+1) \quad = \quad (1-\rho) \cdot \tau_{ij}(t) + \Delta\tau_{ij}(t) \qquad (5.4)$$

The amount of pheromone added to each path after one iteration is inversely proportional to the length of the paths that the ants found (eq. (5.2)). The results for all ants that moved through a path are reflected in the path (eq. (5.3)). Here, A_{ij} is the set of all ants that moved on a path from city i to city j. Negative feedback to avoid local solutions is given as an evaporation coefficient (eq. (5.4)), where the amount of pheromone in the paths, or information from the past, is reduced by a fixed factor (ρ).

The ACO is an effective method to solve the traveling salesman problem (TSP) compared to other search strategies. Table 5.2 shows the optimized values for four benchmark problems and various minima found using other methods (smaller is better, obviously) [31]. The numbers in brackets indicate the number of candidates investigated. The ACO is more suitable for this problem compared to methods such as a genetic algorithm (GA, Section 2.2), simulated annealing (SA), and evolutionary programming (EP). However, it is inferior to the Lee–Kernighan method (the current TSP champion code). The characteristic that specialized methods perform better in static problems is shared by many metaheuristics (high-level strategies which guide an underlying heuristic to increase its performance). Complicated problems, such as TSPs where the distances between cities are asymmetric or where the cities change dynamically, do not have established programs and the ACO is considered to be one of the most promising methods.

The length and pheromone accumulation of paths between cities are stored

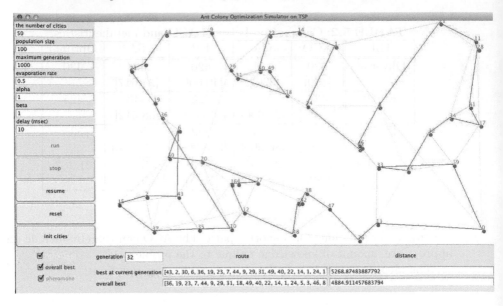

FIGURE 5.6: TSP simulator by ACO.

in a table (Fig. 5.5). Ants can recognize information on their surroundings, and probabilistically decide the next city to visit. The amount of pheromones added to each path after every cycle is inversely proportional to the length of the cycle path.

An applet to solve the TSP using ants is provided for self-study (Fig. 5.6). Operation procedures and displayed contests are almost the same as those with the TSP by a GA (see Section 2.2.7 and Appendix A.3). The paths between cities are color-coded based on the pheromone amount (a darker color means more pheromones). The position of cities can be changed using the GUI; therefore, the effects of pheromones and convergence are easy to understand. Cities can be dynamically changed, yet ants can search with certain accuracy.

5.4 Ant-clustering algorithms

ACO is used in clustering and sorting. The following is a description of ant-clustering.

Ants farm aphids and rear larvae in nests. Livestock and larvae are spatially categorized and placed by size in "farms" and "nurseries" in ant nests. It is considered that this ecology was developed to make feeding work more efficient.

The above behavior can be described by a distributed control model for the probability that an agent picks up (P_{pick}) or drops (P_{drop}) an object.

$$P_{\text{pick}} = (1 - \chi) \cdot \left(\frac{k_{\text{pick}}}{k_{\text{pick}} + f(i)} \right)^2 \qquad (5.5)$$

$$P_{\text{drop}} = \chi \cdot \left(\frac{f(i)}{k_{\text{drop}} + f(i)} \right)^2 \qquad (5.6)$$

Here, $f(i)$ is the density of nearby objects similar to i. To be more precise, this is defined as a function that decreases with an increasing number of similar objects nearby.

$$f(i) = \begin{cases} \frac{\sum_{j \in N(i)} d(i,j)}{|N(i)|} & \text{if } N(i) \neq \phi \\ 1 & \text{if } N(i) = \phi \end{cases} \qquad (5.7)$$

$d(i,j)$ is the similarity (distance in feature space) between objects i and j, and $N(i)$ is a set of neighbors of object i. $d(i,j)$ is normalized such that $0 \leq d(i,j) \leq 1$. $d(i,j) = 0$ means that two objects are identical. Values k_{pick} and k_{drop} are parameters that indicate thresholds in picking up and dropping behavior, respectively. χ is a reaction coefficient determined by the number of objects n in the Moore neighborhood (neighbors in eight directions: top and bottom, right and left, four diagonals).

$$\chi = \frac{n^2}{n^2 + k_{\text{crowd}}^2} \qquad (5.8)$$

k_{crowd} is the threshold, and $\chi = 1/2$ when $n = k_{\text{crowd}}$. χ approaches 1 when there are more objects in the neighborhood, and the behavior is biased towards dropping. On the contrary, fewer objects mean a higher likelihood of picking up behavior.

Equations (5.5) and (5.6) represent a model where agents (ants) randomly acting on a plane move objects based on environmental parameters. In other words, the probability that an agent picks up or drops an object is determined by the density of similar objects in the neighborhood. Agents pick up objects in low-density neighborhoods and drop objects in high-density neighborhoods while moving randomly, and as a result, clusters of similar features are segregated.

The result of ant-clustering, where ants pick up or drop an object based on the above rules and move one step in a randomly chosen direction, is shown in Fig. 5.7. Here, objects of a different color belong to a different class. This two-dimensional space has a torus topology (the right edge is connected to the left edge, and the top edge is connected to the bottom edge). Ants move randomly at first, and then start to form small pre-clusters. These pre-clusters gradually attract similar objects to form larger clusters. Transport by ants

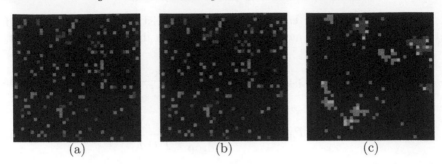

(a)   (b)   (c)

FIGURE 5.7: Ant-clustering.

and accumulation of objects form a positive feedback loop that increases the cluster size, resulting in a clusterization process.

Details on ant-clustering are given in references [28, 45].

5.5   Swarm-based simulation of ant-clustering

The state of execution of ant-clustering by ACO is shown in Fig. 5.8. In this simulation, variables k_{pick}, k_{drop}, k_{crowd}, etc., in Equations (5.5), (5.6), (5.8), number of ants, and number of objects (number of red and blue objects) can be set during execution. Since the "seed" of a random number can be set, re-execution with a different random number is also possible.

Feature values of an object are defined by the `buildObjects` method in the `ModelSwarm.java` file as follows:

```
// Feature vector of red object
v = new double[] { 0.1 + rGen.nextDouble()*0.15,
                   0.1 + rGen.nextDouble()*0.15,
                   0.1 + rGen.nextDouble()*0.15};

// Feature vector of blue object
v = new double[] { 0.7 + rGen.nextDouble()*0.15,
                   0.7 + rGen.nextDouble()*0.15,
                   0.7 + rGen.nextDouble()*0.15};
```

Features in this case are a three-dimensional vector. For generating those objects, a uniform random number vector (each element in the range of 0.0–0.15) is added to $(0.1, 0.1, 0.1)$ or $(0.7, 0.7, 0.7)$. Similarity between two objects is calculated by using the norm of the difference of those feature vectors (the `calcDistance` method inside `DataUnit.java`).

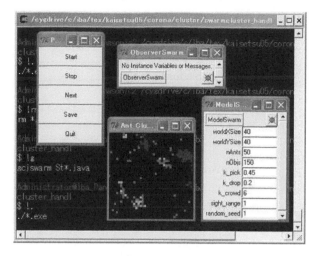

FIGURE 5.8: Ant-clustering with Swarm.

Ant-clustering can be applied to more practical problems by extending this part.

5.6 Ant colony–based approach to the network routing problem

The Internet is one of the most extensively engineered objects today. Packet switching, a method to transfer data developed in the 1960s, breaks down data into small packets, sends these packets to the destination along totally different paths, and recovers the data at the destination. The power of packet switching changed the Internet from an academic and military tool to a mass medium. However, the traffic on the Internet cannot be predicted, and the utilization efficiency is not necessarily high. The routing on the Internet is large-scale and dynamic, so it is not easy to understand the entire structure. Furthermore, throughput and quality of service (QoS) must be increased despite the difficulty of centralized control. The objective in the routing problem is to optimize the routing table (a table specifying the node to which a packet for a predetermined destination should be forwarded) for each node such that the throughput of the entire system is increased (Fig. 5.9).

Each node can only recognize nearby traffic; therefore there is a table for every node. The SPF (shortest path first) routing currently used in the Internet compiles routing tables from information that can be recognized by the respective node. In contrast, Dorigo et al. proposed AntNet routing based

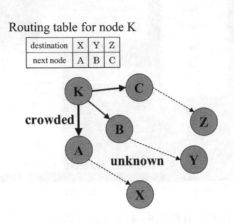

FIGURE 5.9: Routing table.

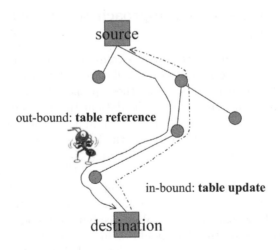

FIGURE 5.10: Agent-based routing.

on ACO. This method deploys software agents on the network that collect routing data by moving back and forth between the source and the destination and updating routing tables in intermediate nodes (Fig. 5.10).

Agent-based routing results in an overhead that is not necessary in static routing; therefore the total traffic and cost performance are important factors.

The steps in AntNet are as follows:

1. Ants are regularly released from each node to random destinations.

2. Ants select paths using pheromones and heuristics and reach their respective destinations. Ants remember the time that they took and the nodes that they visited.

3. Ants that reach their respective destinations return to the origin by moving along the path in reverse order while updating the table in each node.

4. Return to 1.

Similar to the example in the TSP (see eq. (5.5)), the probability that a node is selected as the next node is determined using a weight ω by the following equation:

$$p_{ij}^k(t) = \frac{\omega \cdot \tau_{ij}(t) + (1 - \omega) \cdot \eta_{ij}(t)}{\sum_{h \in J_i^k} \omega \cdot \tau_{ih}(t) + (1 - \omega) \cdot \eta_{ih}(t)}, \tag{5.9}$$

where J_i^k is the set of all cities that ant k in city i can move to (but has not yet visited). The pheromone table in each node is updated through the following equations:

$$\tau_{ij}(t + 1) \quad \Leftarrow \quad (1 - \rho) \cdot \tau_{ij}(t) + \Delta\tau_{ij}(t) \quad \text{(update)} \tag{5.10}$$

$$\Delta\tau_{ij}(t) \quad = \quad \sum_{k=1}^{n} Q(k) \quad \text{(addition)} \tag{5.11}$$

$$\tau_{ij}(t + 1) \quad \Leftarrow \quad \frac{\tau_{ij}(t)}{1 + \Delta\tau_{ij}(t)} \quad \text{(evaporation)} \tag{5.12}$$

The status of the network is not constant in AntNet unlike in ACOs. Therefore, ants move along the paths that packets move to evaluate the status of the network, and return along the same path to reflect the evaluation. As a consequence, the updating of the pheromone tables is not synchronic.

Dorigo et al. tested the AntNet system on a simulator of network file system (NSF), a core network in the USA, and compared it with routing methods such as Bellman–Ford, SPF, and OSPF (open shortest path first) [15]. Figure 5.11 shows the results of tests when there are hot spots. A constant bit rate (CBR) with constant traffic patterns and a variable bit rate (VBR) were

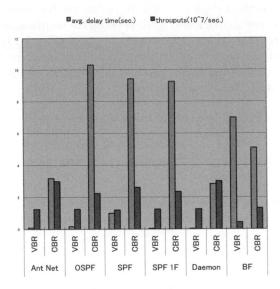

FIGURE 5.11: Comparison of AntNet and other routing methods (from [15]).

tested. The performance under low network load was high with all algorithms, and the throughput when the load was high became similar. In both conditions, AntNet achieved shorter delays with high throughput. In particular, AntNet showed superior performance when the load suddenly changed (hot spots had formed). The overhead of ant agents was negligible in the tests. There are many examples of research on routing using ACOs; for instance, Telecom Bretagne is working on smoothing of QoS. Applications to ATM networks and wireless environments are also being investigated.

5.7   Ant-based job separation

Another well-known ecology of ants is social division of labor. Ant colonies consist of ants with various roles such as queen ants, soldier ants, and worker ants. These roles are called "castes," and some of these (e.g., queen ants) have physically specialized functions. However, general work including collecting and looking after larvae is carried out by worker ants in turn, and no individual is designated to perform a given task. It is also known that for a species where soldier ants divide labor with other worker ants, soldier ants do the work of worker ants if worker ants are removed from the nest. Distributing an appropriate number of ants to each task is necessary to increase the fitness of the colony as a whole. Ants can divide labor without any centralized control; such autonomous distribution of tasks would be useful in the field of robotics or scheduling in factories. The next equation is proposed as a model for assigning workers to multiple tasks through distributed control.

$$T_\theta(s) = \frac{s^n}{s^n + \theta^n} \qquad (5.13)$$

For instance, T is defined as tasks such as feeding larvae. The probability that a given ant does this task T_θ is determined by the amount of pheromone s that the larvae emit and the threshold for each individual θ.

In reality, larvae secrete more pheromone when they are hungry, and reduce the amount of pheromone secretion when caretakers perform their tasks. Individual ants go out to collect food when the amount of detected pheromone becomes lower than a threshold value, and conversely, when the amount of pheromone from larvae exceeds a threshold, ants that returned from collecting food become caretakers. There is a distribution of the threshold in each individual; therefore appropriate numbers of individuals can be distributed to multiple tasks.

Such a behavior model can be applied to the problem of task distribution in robots capable of multiple tasks or fault-tolerant systems. For example, a solution that uses agents called routing wasps has been proposed for scheduling tasks in a factory [16]. In this system, pseudo-pheromones are emitted from tasks in a queue based on priority and wait time. Agents are assigned to each assembly machine, and thresholds to perform specific tasks are determined based on the status of each machine. Agents assign tasks to each machine with probabilities determined by threshold values and amounts of pheromone. Such a system was shown to increase throughput of a factory.

Ant methods are being implemented in various ways in industry. For example, Bios Group[1] based in New Mexico is a consulting firm which builds systems based on swarm intelligence, and has provided methods to make scheduling efficient to Southwest Airlines and P&G, for example. P&G uses distributed scheduling where collaborative decisions such as transport of raw materials and management of factories are made by agents on a network. The swarm approach is used to build a system where the transport path is determined by taking into account the utilization of overcrowded warehouses in the candidate paths.

---

[1]http://www.biosgroup.com/

FIGURE 5.12: A scene of building a living bridge by army ants (photo courtesy of Prof. Salvacion P. Angtuaco [4]).

5.8　Emergent cooperation of army ants

This section presents a multi-agent simulation inspired by army ant behavior. Such cooperation in a multi-agent system can be very valuable for engineering applications. The purpose of this section is to model and comprehend this biological behavior by computer simulation. The following description is mainly based on our previous research results [60].

5.8.1　Altruism of army ants

Altruism refers to behavior that prioritizes benefits to others rather than self and sometimes involves acts of self-sacrifice in order to aid others. Some army ants construct living bridges with their own bodies when they find holes or gullies as obstacles to their marching routes, as shown in [1] (see Fig. 5.12). Such philanthropic acts are different from the regular behavior of the ants, e.g., foraging for and transport of food. However, if more ants participate in bridge construction than is required or if they construct bridges at sites where those are unnecessary, they may actually hamper the food gathering performance of the whole colony. But, in nature, the ants are very keen to balance these actions as per requirements, and it has been confirmed that because of such altruistic activity performance is improved for the group as a whole. In an experiment by Powell and Franks, it was found that the foraging capacity of the army ant colony increased by up to 26% due to this altruistic behavior [98]. In this section, this altruism of ants is modeled and examined in a multi-agent simulation environment (see Fig. 5.13).

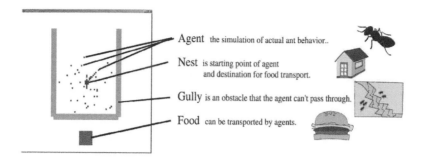

FIGURE 5.13: Simulation environment.

5.8.2 Defining the problem

This section explains the problems handled in the multi-agent simulation. The present simulation serves as a model for the foraging behavior and the altruism of ants. The simulation was performed using the Swarm library. Figure 5.14 shows a screenshot of the simulation screen where an agent represents an ant movement.

The actions include foraging for and transport of food and communications with neighboring ants using pheromone. The nest is the starting point of the agents and also the point to which the agents return with food. The pheromone is released by an agent when it finds food. Just as in nature, once secreted, the pheromone attenuates and disperses, thus disseminating information among the ants about the food locations. A gully hinders movement of agents and fundamentally prevents the agents from passing over it. However, if an agent shows altruism and forms a living bridge over the gully, other agents can pass over the gully. The agents move in accordance with the state transition diagram shown in Fig. 5.15. The behavior of agents in different states is shown in Table 5.3.

The problem is to determine the conditions that induce the transition to the altruism state. But it is not concretely known how ants decide the site and timing of living-bridge construction and when they cease the bridge formation. Therefore, in this section, several hypotheses are proposed as altruism initiation conditions, and experiments were performed for verification.

5.8.3 Judgment criteria for entering the altruism state

5.8.3.1 Hypotheses

Two hypotheses have been proposed as the judgment criteria for altruistic activity by army ants.

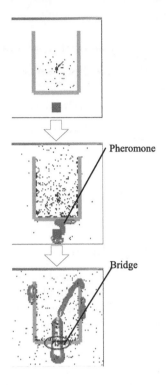

FIGURE 5.14: Swarm-based simulation of army ants.

Model 1: Based on the presence of neighboring ants

An ant will start the formation of a living bridge over a gully only when neighboring ants are present. Hypothetically, this approach will be more efficient compared to forming a bridge blindly because when there are neighboring ants the probability is high that they will utilize the shortcut.

Model 2: Based on the presence of pheromone

As described earlier, agents secrete pheromone when they find food, and this pheromone is used to disseminate information among ants about the location of the food source. The places where pheromone concentrations are higher than a fixed level are the locations that many ants have passed and/or will pass through in the future. Therefore, a living bridge can be formed by judging the pheromone concentration.

In both models, agents leave the bridge after a fixed amount of time passes. We actually used fixed properties optimized by genetic algorithms (shown in

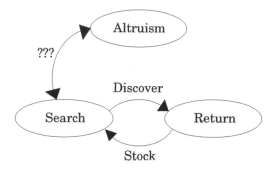

FIGURE 5.15: State transition of agents.

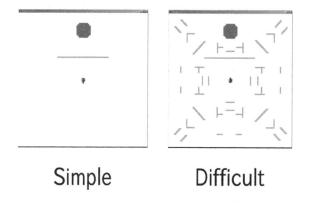

FIGURE 5.16: Maps for experiment.

Table 5.4, [60]). In order to judge their validity, these hypotheses were fed into the simulation and their usefulness was verified empirically.

5.8.3.2 Experiment to verify the hypotheses

The two scenarios shown in Fig. 5.16 were used in the experiment. In these experiments, performance was measured using the number of food items collected within a fixed period of time. Each experiment was repeated 10 times with 20 to 180 agents, increased by 20 at a time, and the mean values were compared.

The experimental results from the simple map are shown in Fig. 5.17. The numbers of agents is shown along the horizontal axis and the number of food items collected within a fixed amount of time is shown along the vertical axis. In the simple map, Model 1 showed slightly higher performance, but

TABLE 5.3: States and behaviors of agents.

State	Behavior
Search	This is the initial condition of the agent and it continues random work until food is found. When food is found, there is a transition to the Return state. Transition to the Altruism state is also possible under "certain" conditions. When pheromone is sensed, the ants are drawn to the higher concentrations.
Return	The food is returned to the nest. In this state the agent moves toward the nest secreting pheromone. After reaching the nest, the agent transits to the Search state. An agent in the Return state knows the position of the nest.
Altruism	A bridge is constructed across the gully. While in this state, movement is impossible for an agent. When certain conditions are met, the bridge is abandoned and the agents transit to the Search state.

TABLE 5.4: Properties used in Models 1 and 2.

	Model 1	Model 2
Number of Steps	700	700
Time	10	150
Radius	2	–
Pheromone Threshhold	–	30

the differences were small and almost no difference in overall efficiency was observed.

Experimental results using the difficult map are shown in Fig. 5.18 and Fig. 5.19. On the whole, Model 2 performed better in the difficult map. Figure 5.19 shows experimental observations for the difficult map on a different scale. Just as before, the horizontal axis represents the number of agents; however, the vertical axis represents the ratio of the total number of times agents crossed bridges to the total number of times agents helped to form bridges. This ratio indicates how useful the bridges formed were. From the data, it was found that Model 2 yielded higher values than Model 1. For Model 1, the ratio was usually about one. This means that even though a bridge was formed, neighboring agents would not have used it efficiently. This was because in the difficult map, unlike the simple map, gullies were present at various locations, causing bridges to be formed at unnecessary sites with Model 1. With Model 2 higher ratios were found compared to that found with Model 1. Although it is not evident from the graph, in Model 2 the bridges were formed only at those sites that were necessary for bringing food to the nest. This was because the pheromone was secreted along the way from the food source to the nest. The concentration of pheromone indicated the optimal sites for bridge construc-

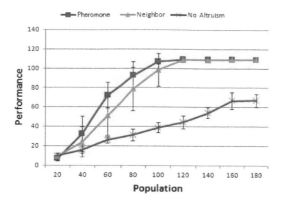

FIGURE 5.17: Simple map – experimental results.

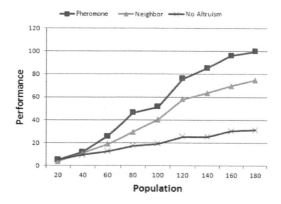

FIGURE 5.18: Difficult map – experimental results 1.

tion. Hence, both the timing and sites of bridge construction were superior in Model 2. However, Model 2 suffers from the drawback that bridges cannot be formed until the foraging sites have been found. In nature, cases are also observed where bridges are formed at necessary sites before foraging sites are found. Thus, for altruistic activity like bridge formation, ants may use the pheromone method along with some other judgment criteria such as the one stated in Model 1.

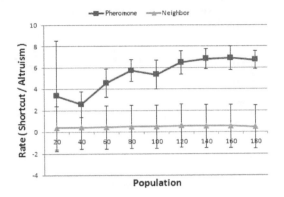

FIGURE 5.19: Difficult map – experimental results 2.

5.8.4   Judgment criteria with reference to chain formation

5.8.4.1   What is chain formation?

Chain formation is another philanthropic cooperative behavior similar to bridge formation. Chains in this case refer to structures formed by the bodies of the ants when the ants encounter extreme differences in heights during their marches. In this way, it is possible for other ants to move safely from one height to another. In their research, Lioni et al. [80] observed the chain formation behavior of ants in nests installed in the laboratory. The results showed that the probability of participation in chain formation P_e and the probability of abandoning chain formation P_s can be approximated by the following equations:

$$P_e = C_{e0} + \frac{C_{e1}X}{C_{e2} + X} \quad (5.14) \qquad P_s = C_{s0} + \frac{C_{s1}X}{C_{s2} + X^\nu}, \quad (5.15)$$

where X is the number of ants participating in chain formation and the other numbers are constants. According to these equations, if many ants are contained in the formed chain then it is easier for them to participate in chain formation but more difficult for them to stop.

Using these formulas as judgment criteria for chain formation, an experiment was conducted to verify the proposed hypotheses, as discussed in the next sections.

5.8.4.2   Experiment to verify the chain formation system

To justify the proposed model of pheromone concentration as the criteria for transition to the altruism state, a comparative study was performed with Lioni's model of chain formation.

In Fig. 5.20, the number of agents is shown on the horizontal axis and

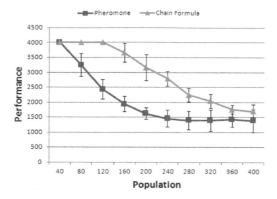

FIGURE 5.20: Performance comparison in terms of foraging time.

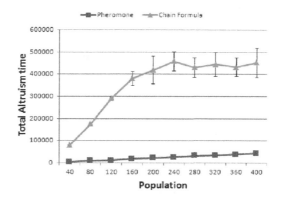

FIGURE 5.21: Performance comparison in terms of altruistic activity.

the time until the completion of foraging on the vertical axis. It was found that for some population sizes, when pheromone concentration is used as the judgment criteria, foraging takes a shorter time than that required for chain formula.

In Fig. 5.21, the number of agents is shown on the horizontal axis and the cumulative time during which the agents are engaged in altruistic behavior on the vertical axis. It was also observed that when pheromone concentration is used as the judgment criterion, the total time during which the agents are engaged in altruistic activity is shorter and is less affected by the population size. On the other hand, when the formulas of Lioni et al. are applied, the time engaged in altruistic behavior increases with the number of agents. Fig. 5.22 compares another aspect of the models. When pheromone concentration was used as the judgment criterion, bridges were constructed at the required sites,

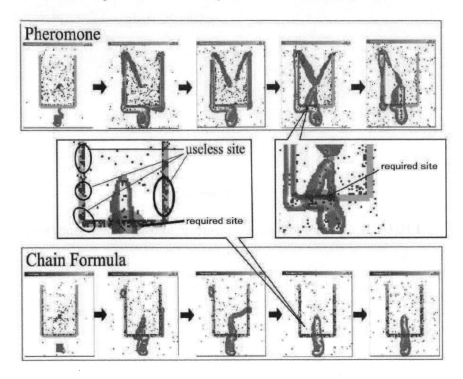

FIGURE 5.22: Comparison of bridge construction sites.

but when the formulas of Lioni et al. were applied, bridges were constructed at many sites other than the required sites. It is also clear from Fig. 5.22 that with the Lioni et al. model, fewer agents are in the Search state as many of them are in the Altruism state.

Procedures using formula (1) of Lioni et al. featured a higher probability of altruistic behavior at sites where agents are apt to congregate. Therefore, more altruistic behavior is expected to occur close to the foraging site and the nest or in between these sites. In Lioni's model, the altruistic behavior is possible without finding foraging sites and this is an advantage over the proposed model based on pheromone concentrations. Nevertheless, the simulation results showed that in terms of performance, measured as foraging speed, the proposed model was superior to Lioni's model. The possible reason behind this could be that in the Lioni et al. experiment calculations were performed by limiting the chain formation sites to one; hence, their model could not be directly applied to an environment with a series of bridge formation sites as used here. Therefore, in consideration of the biology, etc., of army ants, we need to combine the pheromone concentration based model with other judgment criteria.

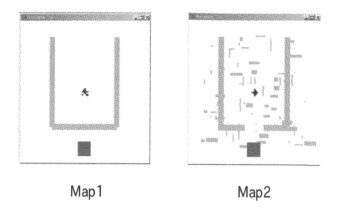

Map 1   Map 2

FIGURE 5.23: Maps used to study the effect of the number of agents.

5.8.5 Changes in strategy based on number of agents

5.8.5.1 Deciding group behavior of army ants

It has been confirmed that the group behavior of army ants is seriously affected by the number of ants that are active [80, 81]. For example, when few ants are available for chain formation, chains are not formed, but when a large number of ants is available, chains are formed at several sites. However, when the number of active ants is moderate, initially several chains are formed. But after a certain time, extension of most of the chains stops and the chains gradually decrease in size; eventually the extension of only one chain continues. However, it is still not clear how the ants count the number of neighboring ants and how this number affects their behavior.

5.8.6 Comparative experiment

In order to monitor the effect of group size on the activity of agents, we performed experiments using the Lioni et al. formulas extended with a minimum limit on group size as an additional condition of chain formation. We compared this scheme with the one that does not take into account the group size. The experiment was performed using two maps, shown in Fig. 5.23. The results of the experiments are shown in Fig. 5.24 and Fig. 5.25. The horizontal axis shows the number of agents, and the vertical axis shows the performance in terms of the number of food items collected within a fixed time. In these figures, "with Check Neighbor" represents the procedure taking the number of neighboring ants into consideration and "without Check Neighbor" indicates the procedure not taking the number of neighboring ants into consideration.

In Map 1, the method that did not take into account information about

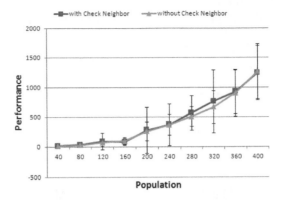

FIGURE 5.24: Effect of neighborhood knowledge (Map 1).

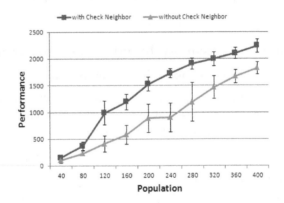

FIGURE 5.25: Effect of neighborhood knowledge (Map 2).

neighboring ants showed high performance. This was because the conditions for bridge formation were relaxed and hence bridges could be formed at an early stage and food could be found easily.

Map 2 was used to investigate whether intelligent behavior can be achieved by avoiding unnecessary bridge formation where a shortcut is not especially necessary for food collection. In this case, better results were obtained with the method that checks the number of neighboring ants.

Figure 5.26 shows how the bridges extend in size with time for Map 1. In the figure, "1st" refers to the largest bridge at the time and "2nd" refers to the next largest bridge. The horizontal axis shows time and the vertical axis shows the two largest bridges. It is apparent from the graph that at first several bridges coexist and extend for about the same length, but finally the differences become greater. Figure 5.27 shows the data obtained in a biological

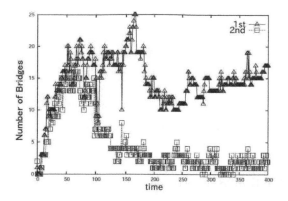

FIGURE 5.26: Changes in size of bridge.

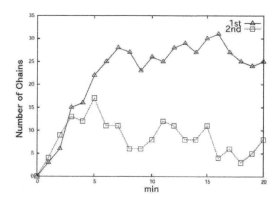

FIGURE 5.27: Changes in size of chain (data plot from [81]).

experiment in the research of Lioni et al. [81]. When chains were formed at two sites, records were kept on how each of the chains extended. In the figure, "1st" and "2nd" show the sizes of the chains at each site.

As can be seen from the two figures, these curves, i.e., simulation data and data from the biological experiment, look quite similar.

5.8.7 Simulation with fixed role assigned

From the previous experiments, it seems that our model has many properties similar to actual army ant behavior. To emphasize the similarity between the simulator agents and the actual army ants, it is important to compare experimental data. We can do that by corresponding the agents' behavior to the army ants' behavior.

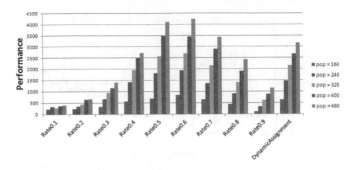

FIGURE 5.28: Experimental results with task assignment.

As a first step, the experiment was performed using a simulator that has agents with a fixed task assigned. Task assignment is one of the signatures that is observed in army ants. Army ants have tasks that depend on someone's rank. Here we consider two different roles for agents in our simulator.

- role A: Search and transport food.

- role B: Build a bridge to support role A.

We carried out the experiment by assigning agents to these two roles with different ratios. Figure 5.28 shows the experimental results where the performance was compared in terms of the number of food items collected within a fixed time. Rate 0.1 means that 10% of the agents were assigned to role B in the simulation. "Dynamic Assignment" labels the experimental results obtained by the simulator used in Section 5.8.5 where the agents have no fixed role.

The results in Fig. 5.28 indicate that a fixed division of roles may be better than a dynamic one. Particularly, Rate 0.5 and Rate 0.6 are better than other ratios. Although "Dynamic Assignment" was not the best, it performed competitively on an average.

This section points out the possibility of role assignment in our simulator. In the real world, it is not possible to know the role assignments of ants to solve this problem. The improvement of Swarm-based simulation with more realistic knowledge is a future research concern.

Chapter 6

Particle Swarm Simulation

> In theory at least, individual members of the school can profit from the discoveries and previous experience of all other members of the school during the search for food. This advantage was documented earlier with reference to bird flocks. It can become decisive, outweighing the disadvantages of competition for food items, whenever the resource is unpredictably disturbed in patches. Thus, larger fish that prey on schools of smaller fish or cephalopods might be expected to hunt in groups for this reason alone [126, p. 442].

6.1 Boids and flocking behaviors

Many scientists have attempted to express the group behavior of flocks of birds and schools of fish, using a variety of methods. Two of the most well-known of these scientists are Reynolds and Heppner, who simulated the movements of birds. Reynolds was fascinated by the beauty of bird flocks [101], and Heppner, a zoologist, had an interest in finding the hidden rules in the instantaneous stops and dispersions of flocks [50]. These two shared a keen understanding of the unpredictable movements of birds; at the microscopic level, the movements were extremely simple, as seen in cellular automata, while at the macroscopic level, the motions were very complicated and appeared chaotic. This is what is called an "emergent property" in the field of Artificial Life (Alife). Their model places a very high weight on the influence of individuals on each other. Similarly, it is known that an "optimum distance" is maintained among individual fish in a fish school (see Fig. 6.1).

This approach is probably not far from the mark as the basis for the social behavior of groups of birds, fish, animals, and, for that matter, human beings. The sociobiologist E.O.Wilson made an interesting suggestion with respect to schools of fish (see the quote at the beginning of this chapter).

As one can understand from this quote, the most useful information to an individual is what is shared from other members of the same group. This hypothesis forms the basis for the particle swarm optimization (PSO) method, which will be explained in Section 6.4.

FIGURE 6.1: Do the movements of a school of fish follow a certain set of rules? (@Coral Sea in 2003)

The collective behavior of a flock of birds emphasized the rules for keeping the optimum distance between an individual and its neighbors.

The computer graphics (CG) video by Reynolds features a group of agents called "boids." Each boid moves according to the sum of three vectors: (1) force to move away from the nearest individual or obstacle, (2) force to move toward the center of the flock, and (3) force to move toward its destination. Adjusting coefficients in this summation results in many behavioral patterns. This technique is often used in special effects and videos in films. Figure 6.2 (simple behavior of flocks) and Fig. 6.3 (situation with obstacles) are examples of simulations of boids.

The following are the details of the algorithm that boids follow. Many individuals (boids) move around in space, and each individual has a velocity vector. The three factors below result in a flock of boids.

1. Avoid collision: attempt to avoid collision with nearby individuals.

2. Match pace: attempt to match the velocity of nearby individuals.

3. Move to center: attempt to be surrounded by nearby individuals.

Each boid has an "optimum distance" to avoid collision, and behaves so as to maintain this distance with its nearest neighbor [102]. Collision becomes a concern if the distance between nearby boids becomes shorter than the "optimum distance." Therefore, to avoid collision, each boid slows down if the nearest boid is ahead and speeds up if the nearest boid is behind (Fig. 6.4).

The "optimum distance" is also used to prevent the risk of straying from the flock. If the distance to the nearest boid is larger than the "optimum distance," each boid speeds up if the nearest boid is ahead and slows down if it is behind (Fig. 6.5).

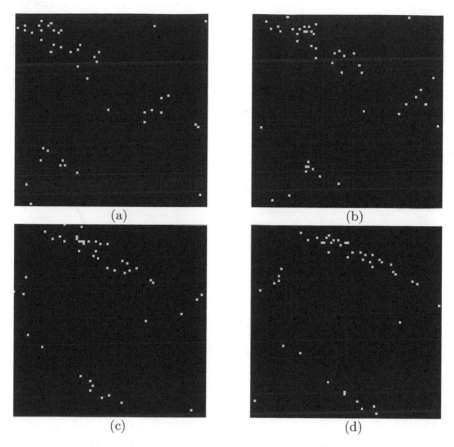

FIGURE 6.2: Simple behavior of boids ((a)⇒(b)⇒(c)⇒(d)).

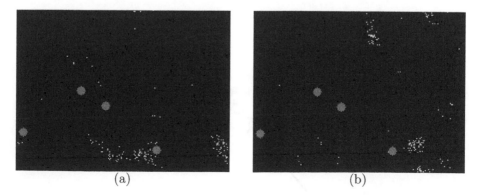

FIGURE 6.3: Boids in a situation with obstacles ((a)⇒(b)).

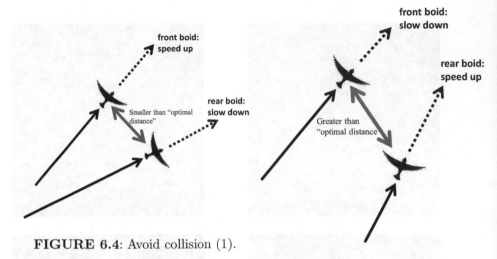

FIGURE 6.4: Avoid collision (1).

FIGURE 6.5: Avoid collision (2).

Here, "ahead" and "behind" are defined as ahead or behind a line that crosses the boid's eyes and is perpendicular to the direction in which the boid is moving (Fig. 6.6). Boids try to move parallel to (with the same vector as) their nearest neighbor. Here, there is no change in speed. Furthermore, boids change velocity so as to always move toward the center of the flock (center of gravity of all boids).

In summary, the velocity vector ($\vec{v_i}(t)$) of the i-th boid is updated at time t as follows (see Fig. 6.7):

$$\vec{v_i}(t) = \vec{v_i}(t-1) + \vec{Next}_i(t-1) + \vec{G}_i(t-1) \tag{6.1}$$

where $\vec{Next}_i(t-1)$ is the velocity vector of the nearest boid to individual

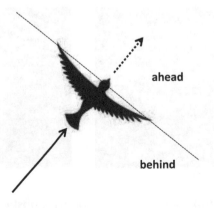

FIGURE 6.6: Ahead or behind a line that crosses the boid's eyes.

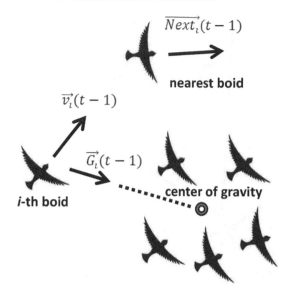

$\overrightarrow{Next_i}(t-1)$

nearest boid

$\vec{v_i}(t-1)$

$\vec{G_i}(t-1)$

i-th boid

center of gravity

FIGURE 6.7: Updating the velocity vector.

i, and $\vec{G_i}(t-1)$ is the vector from individual i to the center of gravity. The velocity at one step before, i.e., $\vec{v_i}(t-1)$, is added to take inertia into account.

Each boid has its own field of view (Fig. 6.8) and considers boids within its view when finding the nearest neighbor. However, the coordinates of all boids, including those out of view, are used to calculate the center of gravity.

Kennedy and Eberhart designed an effective optimization algorithm using the mechanism behind boids [71]. This is called particle swarm optimization (PSO), and numerous applications are reported. The details are provided in Section 6.4.

6.2 Simulating boids with Swarm

As explained in Section 6.1, the velocity update equation ($\vec{v_i}(t)$) of boid i at time t was defined as follows:

$$\vec{v_i}(t) = \vec{v_i}(t-1) + \overrightarrow{Next_i}(t-1) + \vec{G_i}(t-1) \tag{6.2}$$

where $\overrightarrow{Next_i}(t-1)$ is the velocity vector of the boid closest to object i, and $\vec{G_i}(t-1)$ is the vector toward the center of gravity from object i.

In the easiest Boid program, this is achieved by the "step" method of Bug.java as follows:

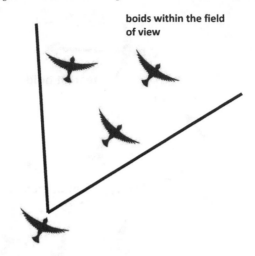

boids within the field of view

FIGURE 6.8: Each boid has its own field of view.

```
// the number of ''boid'' in the field of sight is substituted
//                           in ''Num'' in the lines above these.
gX /= Num;
gY /= Num;
// coordinates of the center of gravity are obtained in (gX, gY).


// in the lines above,
// the coordinates of the closest boid are obtained in (minDx,
   minDy).
Bug nearestBug =
    (Bug)world.getObjectAtX$Y((minDX+worldXSize)%worldXSize,
                      (minDY+worldYSize)%worldYSize);
// the closest boid is kept in nearestBug.

//direction of center of gravity is obtained in this variable
float gVX = 0.0f, gVY = 0.0f;
float tmp = (float)Math.sqrt((float)((gX-xPos)*(gX-xPos)+(gY-
yPos)*(gY-yPos)));
    gVX = (float)( gX - xPos ) / tmp;
    gVY = (float)( gY - yPos ) / tmp;
// (xPos, yPos) shows the current position of boid.

float sVX = 0.0f, sVY = 0.0f; // obtains its own velocity vector
  sVX = (float)Math.cos( direction );
  sVY = (float)Math.sin( direction );
```

```
float nVX = 0.0f, nVY = 0.0f;
// obtains the velocity vector of the closest boid
  nVX = (float)Math.cos( nearestBug.direction );
  nVY = (float)Math.sin( nearestBug.direction );

float fVX, fVY; // obtains the new directional vector
  fVY = gravityWeight * gVY + sVY + nearWeight * nVY;
  newDirection = Vector2Direction( fVX, fVY );

float dX = (float)(minDX - xPos);
float dY = (float)(minDY - yPos);
// from the current position,
// obtains the directional vector towards the closest boid
float inner = dX * fVX + dY * fVY;   // used to determine the
proximity
float nearestDist = (float)Math.sqrt( dX * dX + dY * dY );

// obtains the new speed ''new speed''
//         by changing the current speed ''speed''
if( inner > 0 ){ // If the closest boid was in front of self
    if( nearestDist > optDistance ){
            newSpeed = speed * accel;   // speed up
        else{
            newSpeed= speed / accel;   // speed down
        }
    }else{  // if the closest boid was behind
            if( nearestDist > optDistance ){
                newSpeed= speed / accel;   // speed down
            }else{
                newSpeed = speed * accel; //speed up
        }
    }
}
```

After that, the following lines obtain the new coordinates (`newX`, `newY`) of the boid.

```
newX = xPos + (int)( newSpeed * Math.cos( newDirection ) );
newY = yPos + (int)( newSpeed * Math.sin( newDirection ) );
```

Note that the behavior of the "boid" group slightly differs by changing `gravityWeight` and `nearWeight` to various values. Apart from this, the extended versions of "boid" are also provided as follows:

- Two types of tribes/races of boid are introduced.

- Obstacles are placed in the map.

6.3   Swarm Chemistry

Swarm Chemistry is a system designed by Sayama [104] to model and simulate the behavior of groups of, for example, fish and ants.

Agents in Swarm Chemistry move in two-dimensional space according to simple rules based on a number of parameters. The position vector X_i' and velocity vector V_i' at the next step of the agents in the population are determined from the current position vector X_i and velocity vector V_i according to the following process.

1. Find agents near an agent X_i with Euclidean distance less than r.

2. If there are no agents nearby, a random value of $[-0.5, 0.5]$ is added to the x and y components of the acceleration A_i of X_i.

3. If there is at least one agent nearby, the acceleration A_i of X_i is updated using the following equation. Here, $\bar{X}_i$ is the average position and $\bar{V}_i$ is the average velocity of nearby agents.

$$A_i = c_1(X_i - \bar{X}_i) + c_2(V_i - \bar{V}_i) + c_3\Sigma_{j=1}^{N}(X_i - X_j)/|X_i - X_j|^2 \quad (6.3)$$

In addition, a random value of $[-0.5, 0.5]$ is added to the x and y components of A_i with a probability c_4.

4. Add A_i to V_i'.

5. If $|V_i'| > v_m$, multiply V_i' by a constant value such that the absolute value becomes v_m.

6. Update V_i' using the equation:

$$V_i' \leftarrow c_5(v_n/|V_i'| \cdot V_i') + (1 - c_5)V_i' \quad (6.4)$$

7. Perform the above procedure on all agents.

8. Substitute the velocity V_i of each agent with V_i' to update the velocity.

9. Add V_i to the positions of each agent X_i to obtain the positions at the next step X_i'.

There are eight control parameters in the update process: $r, v_n, v_m, c_1, c_2, c_3,$ c_4, c_5. Using multiple populations with different parameters results in interactions between populations, and complex behavior can be observed.

The populations shown in Fig. 6.9 are an example of behavior arising from this system, where a swarm group rotates around another swarm group. The control parameters are

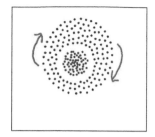

FIGURE 6.9: Behavior in Swarm Chemistry (1).

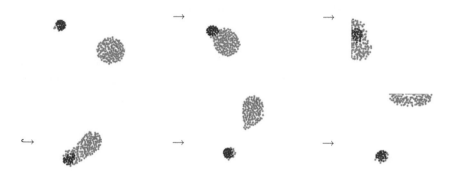

FIGURE 6.10: Behavior in Swarm Chemistry (2).

60*{73.03, 0.61, 5, 0.75, 0.17, 28.81, 0.32, 0.37}
140*{93.28, 5.15, 10.71, 0.64, 0.58, 96.71, 0.07, 0.41}

Where 60 and 140 are the number of boids in each population.

Figure 6.10 shows a behavior where the large swarm group is reflected at the walls and is sometimes attracted to the small swarm group. The control parameters are

164*{52.86, 9.69, 13.19, 0.93, 0.5, 23.84, 0.3, 0.85}
36*{73.31, 0.76, 3.47, 0.35, 0.32, 7.47, 0.09, 0.22}

Figure 6.11 is a screenshot of a system where the control parameters are adjusted through interactive evolutionary computation (IEC) to obtain a system that behaves according user preferences. This system is a Java applet that is openly available online.[1]

---

[1]http://www.iba.t.u-tokyo.ac.jp/~akio/swarm_chemistry.html

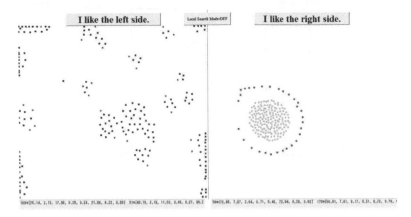

FIGURE 6.11: A snapshot of Swarm Chemistry.

A user observes the behavior of two systems to the right and left, and chooses the system that he likes. Repeating this pairwise comparison optimizes the parameter to what the user likes. When a system very close to what the user wants appears, the user can fine-tune it by clicking the "Local Search Mode" button at the top.

6.4   PSO: particle swarm optimization

This section introduces the optimization method "particle swarm optimization" (PSO), which differs slightly from GA and GP. PSO is an algorithm from the field of Swarm Intelligence. It was first described by Kennedy and Eberhart as an alternative to GA in 1995 [71]. The algorithm for PSO was conceived on the basis of observations of certain social behavior in lower-class animals or insects. In contrast to the concept of modifying genetic codes using genetic operations as used in GA, in PSO the moving individuals (called "particles") are considered where the next movement of an individual is determined by the motion of the individual itself and that of the surrounding individuals. It has been established that PSO has capabilities equal to those of GA for function optimization problems. There have been several comparative studies on PSO and standard GA (see [3, 35, 51, 73]).

Below we describe the origins of PSO, outline the procedure, compare its search efficiency with that of GA, and provide some examples of its application.

6.4.1 PSO algorithm

The classic PSO was intended to be applied to optimization problems. It simulates the motion of a large number of individuals (or "particles") moving in a multi-dimensional space [71]. Each individual stores its own location vector $(\vec{x_i})$, velocity vector $(\vec{v_i})$, and the position at which the individual obtained the highest fitness value $(\vec{p_i})$. All individuals also share information regarding the position with the highest fitness value for the group $(\vec{p_g})$.

As generations progress, the velocity of each individual is updated using the best overall location obtained up to the current time for the entire group and the best locations obtained up to the current time for that individual. This update is performed using the following formula:

$$\vec{v_i} = \chi(\omega\vec{v_i} + \phi_1 \cdot (\vec{p_i} - \vec{x_i}) + \phi_2 \cdot (\vec{p_g} - \vec{x_i})) \tag{6.5}$$

The coefficients employed here are the convergence coefficient χ (a random value between 0.9 and 1.0) and the attenuation coefficient ω, while ϕ_1 and ϕ_2 are random values unique to each individual and the dimension, with a maximum value of 2. When the calculated velocity exceeds some limit, it is replaced by a maximum velocity V_{max}. This procedure allows us to hold the individuals within the search region during the search.

The locations of each of the individuals are updated at each generation by the following formula:

$$\vec{x_i} = \vec{x_i} + \vec{v_i} \tag{6.6}$$

The overall flow of the PSO is as shown in Fig. 6.12. Let us now consider the specific movements of each individual (see Fig. 6.13). A flock consisting of a number of birds is assumed to be in flight. We focus on one of the individuals (Step 1). In the figure, the ◯ symbols and linking line segments indicate the positions and paths of the bird. The nearby ◉ symbol (on its path) indicates the position with the highest fitness value on the individual's path (Step 2). The distant ◉ symbol (on the other bird's path) marks the position with the highest fitness value for the flock (Step 2). One would expect that the next state will be reached in the direction shown by the arrows in Step 3. Vector ① shows the direction followed in the previous steps; vector ② is directed toward the position with the highest fitness for the flock; and vector ③ points to the location where the individual obtained its highest fitness value so far. Thus, all these vectors, ①, ②, and ③, in Step 3 are summed to obtain the actual direction of movement in the subsequent step (see Step 4).

A simulator is available for investigating the PSO search process. Figure 6.14 is a screenshot of the simulator. Interested readers are referred to Appendix A.2 for a detailed description of how the simulator is operated.

The efficiency of this type of PSO search is certainly high because focused searching is available near optimal solutions in a relatively simple search space. However, the canonical PSO algorithm often gets trapped in a local optimum in multimodal problems. Because of that, some sort of adaptation is necessary in order to apply PSO to problems with multiple sharp peaks.

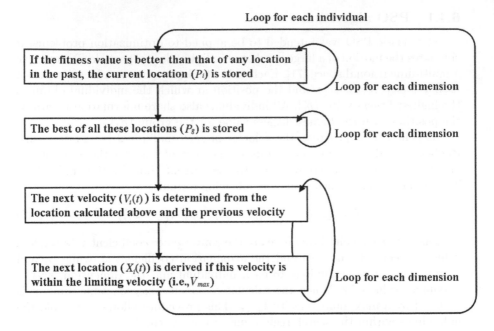

FIGURE 6.12: Flow chart of the PSO algorithm.

To overcome the above limitation, a GA-like mutation can be integrated with PSO [51]. This hybrid PSO does not follow the process by which every individual of the simple PSO moves to another position inside the search area with a predetermined probability without being affected by other individuals, but leaves a certain ambiguity in the transition to the next generation due to Gaussian mutation. This technique employs the following equation:

$$mut(x) = x \times (1 + Gaussian(\sigma)), \qquad (6.7)$$

where σ is set to be 0.1 times the length of the search space in one dimension. The individuals are selected at a predetermined probability and their positions are determined at the probability under the Gaussian distribution. Wide-ranging searches are possible at the initial search stage and search efficiency is improved at the middle and final stages by gradually reducing the appearance ratio of the Gaussian mutation at the initial stage. Figure 6.15 shows the PSO search process with a Gaussian mutation. In the figure, V_{lbest} represents the velocity based on the local best, i.e., $\vec{p_i} - \vec{x_i}$ in eq. (6.5), whereas V_{gbest} represents the velocity based on the global best, i.e., $\vec{p_g} - \vec{x_i}$.

6.4.2   Comparison with GA

Let us turn to a comparison of the performance of PSO with that of the GA using benchmark functions to examine the effectiveness of PSO.

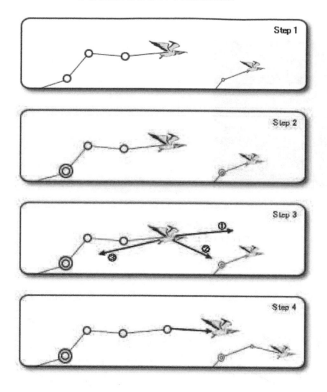

FIGURE 6.13: In which way do birds fly?

For the comparison, $F8$ (Rastrigin's function) and $F9$ (Griewangk's function) are employed. These are defined as:

$$F8(x_1, x_2) = 20 + x_1^2 - 10\cos(2\pi x_1) + x_2^2 - 10\cos(2\pi x_2)$$
$$-(-5.11 \leq x_i \leq 5.11)$$
$$F9(x_1, x_2) = \frac{1}{4000}\sum_{i=1}^{2}(x_i - 100)^2 - \prod_{i=1}^{2}\cos\left(\frac{x_i - 100}{\sqrt{i}}\right) + 1(-10 \leq x_i \leq 10)$$

Figures 6.16 and 6.17 show the shapes of $F8$ and $F9$, respectively. $F8$ and $F9$ seek the minimum value. $F8$ contains a large number of peaks so that its optimization is particularly difficult.

Comparative experiments were conducted with PSO and GA using the above benchmark functions. PSO and GA were repeatedly run 100 times. Search space ranges for the experiments are listed in Table 6.1. PSO and GA parameters are given in Table 6.2.

The performance results are shown in Figs. 6.18 and 6.19, which plot the fitness values against the generations. Table 6.3 shows the averaged best fitness values over 100 runs. As can be seen from the table and the figures, the

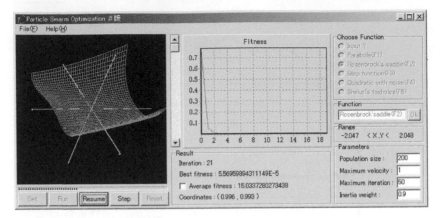

FIGURE 6.14: PSO simulator.

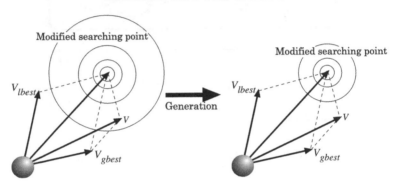

FIGURE 6.15: Concept of searching process by PSO with a Gaussian mutation.

combination of PSO with a Gaussian mutation allows us to achieve a performance that is almost equal to that of the canonical PSO for the unimodals, and a better performance than the canonical PSO for the multimodals. The experimental results with other benchmark functions are further discussed in [58].

PSO is a stochastic search method, as are GA and GP, and its method of adjustment of $\vec{p}_i$ and $\vec{p}_g$ resembles crossover in GA. It also employs the concept of fitness, as in evolutionary computation. Thus, the PSO algorithm is strongly related to evolutionary computation (EC) methods. In conceptual terms, one could place PSO somewhere between GA and EP.

However, PSO has certain characteristics that other EC techniques do not have. GA operators directly operate on the search points in a multi-dimensional search space, while PSO operates on the motion vectors of particles which in turn update the search points (i.e., particle positions). In other words, GA operators are position specific and PSO operators are direction

TABLE 6.1: Search space for test functions.

Function	Search space
$F8$	$-65.535 \leq x_i < 65.536$
$F9$	$-10 \leq x_i \leq 10$

TABLE 6.2: PSO and GA parameters.

Parameter	PSO, PSO with Gaussian	Real-valued GA
Population	200	200
V_{max}	1	
Generation	50	50
ϕ_1, ϕ_2	upper limits $= 2.0$	
Inertia weight	0.9	
Crossover ratio		0.7(BLX-α)
Mutation	0.01	0.01
Elite		0.05
Selection		tournament (size=6)

TABLE 6.3: Average best fitness of 100 runs for experiments.

	Gen	GA	PSO	PSO with Gaussian
$F8$	1	4.290568	3.936564	3.913959
	10	0.05674	0.16096	0.057193
	20	0.003755	0.052005	0.002797
	30	0.001759	0.037106	0.000454
	40	0.001226	0.029099	0.000113
	50	0.000916	0.02492	3.61E-05
$F9$	1	0.018524	0.015017	0.019726
	10	0.000161	0.000484	0.000145
	20	1.02E-05	0.000118	1.43E-05
	30	3.87E-06	6.54E-05	4.92E-06
	40	2.55E-06	5.50E-05	2.04E-06
	50	1.93E-06	4.95E-05	1.00E-06

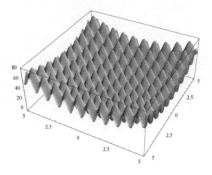

FIGURE 6.16: Rastrigin's function ($F8$).

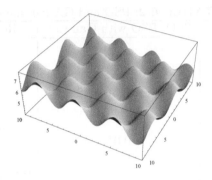

FIGURE 6.17: Griewangk's function ($F9$).

specific. One of the reasons PSO has gathered so much attention is the tendency of its individuals to proceed directly toward the target. This feature resembles the behavior of BUGS, which is described in Section 6.6.

In the chapter "The Optimal Allocation of Trials" in his book [52], Holland ascribes the success of EC to the balance of "exploitation," through search of known regions, with "exploration," through search, at finite risks, of unknown regions. PSO is adept at managing such subtle balances. These stochastic factors enable PSO to make thorough searches of the relatively promising regions and, due to the momentum of speed, also allows effective searches of unknown regions. Theoretical research is currently underway to derive optimized values for PSO parameters by mathematical analysis, for stability and convergence (see [18, 72]).

6.4.3   Examples of PSO applications

PSO has been applied to an analysis of trembling of the human body [34]. Trembling has two types, ordinary shivering and the type of shaking that is

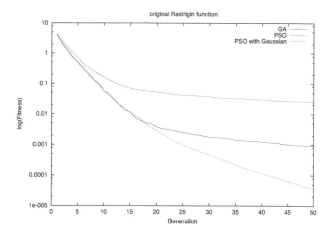

FIGURE 6.18: Standard PSO versus PSO with a Gaussian mutation for $F8$.

caused by Parkinson's disease or other illnesses. The authors used a combination of PSO and a neural network to distinguish between the types. The sigmoid function given below was optimized with PSO in a layered network with 60 input units, 12 hidden nodes, and 2 output units, thus:

$$output = \frac{1}{1 + e^{-k \sum w_i x_i}},$$

where x_i and w_i were the inputs and weights to each of the hidden layers and output layers, respectively. Optimization of the weight indirectly causes changes in the network structure. Ten healthy controls and twelve patients took part in this experiment. The system succeeded in distinguishing correctly between the types of shaking in the subjects with 100% accuracy.

PSO has been applied to problems of electric power networks [86]. In their research, the experiments were conducted employing selection procedures that were effective for standard PSO and an extended version (EPSO) with a self-adaptive feature. The problem of "losses" in electric power networks refers to searching out the series of control actions needed to minimize power losses. The objective function for this included the level of excitation of generators and adjustments to the connections to transformers and condensers, i.e., the control variables included both continuous and discrete types. The maximum power flow and the permitted voltage level were imposed as boundary conditions, and the algorithm searched for the solution with the minimum loss. Miranda and Fonseca [81] conducted a comparative experiment with EPSO and simulated annealing (SA), conducting 270 runs in each system and comparing the mean of the results. EPSO rapidly identified a solution that was close to the optimal one. SA converged more slowly. Comparison of the mean square errors indicated that SA did not have as high a probability of arriving at the optimal solution as EPSO. PSO has also been successfully applied

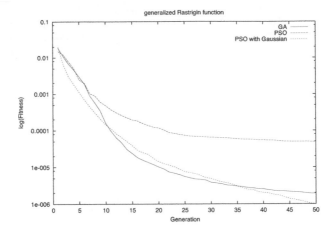

FIGURE 6.19: Standard PSO versus PSO with a Gaussian mutation for $F9$.

to the economic load dispatch (ELD) problem for least cost power generation [42, 95]. These findings indicate that PSO can be trusted as a sufficiently robust method for solving real problems.

Practical research has also been conducted applying PSO to optimize the mixing of materials for the production of valuable excretions by microorganisms [72]. The authors compared PSO with traditional methods of experimental design, finding that the mixture indicated by PSO resulted in more than a doubling of performance. When materials of low quality were used, the search efficiency was quite poor in the initial stages, but ultimately, PSO provided superior results. These findings confirmed that PSO offers good robustness against changes in the environment.

6.5   ABC algorithm

Bees, along with ants, are well-known examples of social insects (Fig. 6.20). Bees are classified into three types: employed bees, onlooker bees, and scout bees. Employed bees fly in the vicinity of feeding sites they have identified, sending information about food to onlooker bees. Onlooker bees use the information from employed bees to perform selective searches for the best food sources from the feeding site. When information about a feeding site is not updated for a given period of time, its employed bees abandon it and become scout bees that search for a new feeding site. The objective of a bee colony is to find the highest-rated feeding sites. The population is approximately half employed bees and scout bees (about 10–15% of the total); the rest are onlooker bees.

FIGURE 6.20: A bee colony.

The waggle dance (a series of movements) performed by employed bees to transmit information to onlooker bees is well known (Fig. 6.21). The dance involves shaking the hindquarters and indicating the angle with which the sun will be positioned when flying straight to the food source, with the sun represented as straight up. For example, a waggle dance performed horizontally and to the right with respect to the nest combs means "fly with the sun at 90 degrees to the left." The speed of shaking the rear indicates the distance to the food; when the rear is shaken quickly, the food source is very near, and when shaken slowly it is far away. Communication via similar dances is also performed with regard to pollen and water collection, as well as the selection of locations for new hives.

The Artificial Bee Colony (ABC) algorithm [65, 66], initially proposed by Karaboga et al., is a swarm optimization algorithm that mimics the foraging behavior of honey bees. Since ABC was designed, it has been proved that ABC, with fewer control parameters, is very effective and competitive with other search techniques such as Genetic Algorithm (GA), Particle Swarm Optimization (PSO), and Differential Evolution (DE).

In ABC algorithms, an artificial swarm is divided into employed bees, onlooker bees, and scouts. N d-dimensional solutions to the problem are randomly initialized in the domain and referred to as food sources. Each employed bee is assigned to a specific food source x_i and searches for a new food source

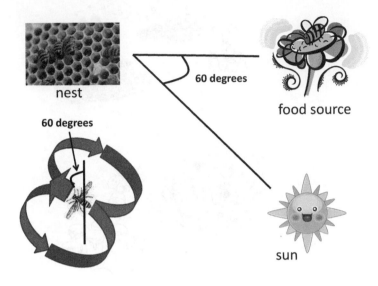

FIGURE 6.21: Waggle dance.

v_i by using the following operator:

$$v_{ij} = x_{ij} + \text{rand}(-1, 1) \times (x_{ij} - x_{kj}), \qquad (6.8)$$

where $k \in \{1, 2, \cdots, N\}$, $k \neq i$, and $j \in \{1, 2, \cdots, d\}$ are randomly chosen indices. v_{ij} is the jth element of the vector v_i. If the trail to a food source is outside of the domain, it is reset to an acceptable value. The v_i obtained is then evaluated and put into competition with x_i for survival. The bee prefers the better food source. Unlike employed bees, each onlooker bee chooses a preferable source according to the food source's fitness to do further searches in the food space using eq. (6.8). This preference scheme is based on the fitness feedback information from employed bees. In classic ABC [65], the probability of the food source x_i that can be exploited is expressed as

$$p_i = \frac{fit_i}{\sum_{j=1}^{N} fit_j}, \qquad (6.9)$$

where fit_i is the fitness of the ith food source, x_i. For the sake of simplicity, we assume that the fitness value is non-negative and that the larger, the better. If the trail v_i is superior to x_i in terms of profitability, this onlooker bee informs the relevant employed bee associated with the ith food source,

x_i, to renew its memory and forget the old one. If a food source cannot be improved upon within a predetermined number of iterations, defined as `Limit`, this food source is abandoned. The bee that was exploiting this food site becomes a scout and associates itself with a new food site that is chosen via some principle. In canonical ABC [65], the scout looks for a new food site by random initialization.

The details of the ABC algorithm are described below. The pseudocode of the algorithm is shown in **Algorithm 1**.

Step 0: Preparation The total number of search points (N), total number of trips (T_{max}), and a scout control parameter (`Limit`) are initialized. The numbers of employed bees and onlooker bees are set to be the same as the total number of search points (N). The value of the objective function f is taken to be non-negative, with larger values being better.

Step 1: Initialization 1 The trip counter k is set to 1, and the number of search point updates s_i is set to 0. The initial position vector for each search point $x_i = (x_{i1}, x_{i2}, x_{i3}, \cdots, x_{id})^T$ is assigned random values. Here, the subscript i ($i = 1, \cdots, N$) is the index of the search point, and d is the number of dimensions in the search space.

Step 2: Initialization 2 Determine the initial best solution *best*.

$$i_g = \operatorname*{argmax}_i f(x_i) \tag{6.10}$$

$$\textbf{\textit{best}} = x_{i_g} \tag{6.11}$$

Step 3: Employed bee search The following equation is used to calculate a new position vector v_{ij} from the current position vector x_{ij}.

$$v_{ij} = x_{ij} + \phi \cdot (x_{ij} - x_{kj}) \tag{6.12}$$

Here, j is a randomly chosen dimensional number, k is the index for some randomly chosen search point other than i, and ϕ is a uniform random number in the range $[-1, 1]$. The position vector x_i and the number of search point updates s_i are determined according to the following equation:

$$I = \{i \mid f(x_i) < f(v_i)\} \tag{6.13}$$

$$x_i = \begin{cases} v_i & i \in I \\ x_i & i \notin I \end{cases} \tag{6.14}$$

$$s_i = \begin{cases} 0 & i \in I \\ s_i + 1 & i \notin I \end{cases} \tag{6.15}$$

Step 4: Onlooker bee search The following two steps are performed.

Algorithm 1 The ABC algorithm

Require: T_{max}, `#. of employed bees` (=`No. of onlooker bees`),`Limit`
  Initialize food sources
  Evaluate food sources
  $i = 1$
  while $i < T_{max}$ **do**
    Use employed bees to produce new solutions
    Evaluate the new solutions and apply greedy selection process
    Calculate the probability values using fitness values
    Use onlooker bees to produce new solutions
    Evaluate new solutions and apply greedy selection process
    Determine abandoned solutions and use scouts to generate new ones randomly
    Remember the best solution found so far
    $i = i + 1$
  end while
  Return best solution

---

1. Relative ranking of search points
   The relative probability P_i is calculated from the fitness fit_i, which is based on the evaluation score of each search point. Note that $fit_i = f(\boldsymbol{x}_i)$. The onlooker bee search counter l is set to 1.

$$P_i \quad = \quad \frac{fit_i}{\sum_{j=1}^{N} fit_j} \tag{6.16}$$

2. Roulette selection and search point updating
   Search points are selected for updating based on the probability P_i, calculated above. After search points have been selected, perform a procedure as in **Step 3** to update the search point position vectors. Then, let $l = l + 1$ and repeat until $l = N$.

Step 5: Scout bee search   Given a search point for which $s_i \geq$ `Limit`, random numbers are used to exchange generated search points.

Step 6: Update best solution   Update the best solution *best*.

$$i_g \quad = \quad \operatorname*{argmax}_{i} f(\boldsymbol{x}_i) \tag{6.17}$$

$$\boldsymbol{best} \quad = \quad \boldsymbol{x}_{i_g} \text{ when } f(x_{i_g}) > f(\boldsymbol{best}) \tag{6.18}$$

$$\tag{6.19}$$

Step 7: End determination   End if $k = T_{max}$. Otherwise, let $k = k + 1$ and return to **Step 3**.

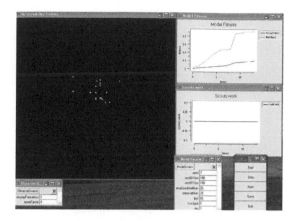

FIGURE 6.22: ABC simulator with Swarm.

ABC has recently been improved in many aspects. For instance, we analyzed the mechanism of ABC to show a possible drawback of using parameter perturbation. To overcome this deficiency, we have proposed a new nonseparable operator and embedded it in the main framework of the cooperation mechanism of bee foraging (see [47] for details).

A swarm-based ABC simulation has been provided (Fig. 6.22). In this implementation of ABC, optimization is performed using CEC2005 benchmark functions [110] as the objective function set. Figure 6.23 shows an example of optimizing the F_8 Shifted Rotated Ackley's function. When scout bees find a local solution one can see them actively approaching an optimal solution. In the window on the left, the movement of employed bees will be in green, that of onlooker bees will be in yellow, and that of scouts will be drawn in red. The window to the upper right shows the average and maximum fitness values by generation. The window in the bottom right shows the number of scout bees over time.

Control of ABC is as follows (Fig. 6.22).

- Start button
  Starts the simulation.

- Stop button
  Pauses the simulation.

- Next button
  Advances the simulation one time step.

- Save button
  Not used in this simulation.

- Quit button
  Terminates the simulation.

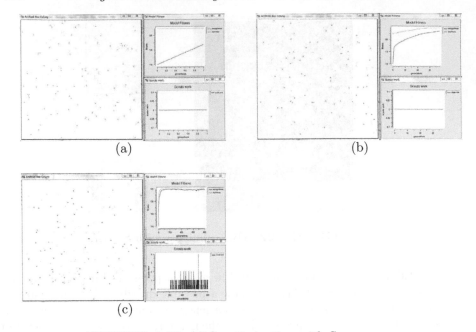

FIGURE 6.23: ABC optimization with Swarm.

- `ObserverSwarm` parameter probe
  - `displayFrequency`: Screen refresh frequency
  - `zoomFactor`: Display zoom factor
- `ModelSwarm` parameter probe
  - `seed`: Random number generator seed
  - `worldXSize,worldYSize`: ABC screen width
  - `beeNum`: Number of bees (= number of search points)
  - `limit`: Scout bee activation parameter
  - `tmax`: Maximum number of generations. When `tmax`= 0, the simulation will continue to run until "stop" is pressed.
  - `functype`: Benchmark function settings
  - `dim`: Dimensions in the simulation space

`ModelSwarm` parameters can be changed before the simulation starts (before pressing "Start" or "Next"). Note that changed parameters will not take effect until the Enter key is pressed after input.

The objective functions implemented are F_1–F_{12} of the CEC2005 benchmark functions [110] (see also Fig. A.2; note that these functions are differently numbered). Setting the parameter functype to a value 1–12 will set the associated function as the objective function.

An arbitrary function can be defined as the objective function. For this purpose, after setting the desired function, you should proceed as follows:

1. Change the following parts of `ModelSwarm.java` (the `setFunc` and `func` functions) and make appropriate definitions under case 0.

```
/* Define range */
void setFunc(){
....
case 0:
xMax= (define this here; range will be [-xMax,xMax])
break;
....

....
/* Define function */
double func(double po[]){
....
case 0:
result = (The function result;
            point coordinates are stored in po[], dimensions
            in dim)
break;
....
```

2. After recompiling, execute as `functype:0`.

It is also possible to change the dimensions and domain of the function. Note that the display in the left Swarm window is for two dimensions.

6.6   BUGS: a bug-based search strategy

This section describes another new approach to strategic learning, based on the concept of swarm intelligence. Simple GAs optimize functions by adaptive combination (crossover) of coded solutions to problems (i.e., points in the problem's search space). In this approach, an analogy is made between the value (at a given point) of a function to be maximized and the density of bacteria at that point. More specifically, the adaptive learning used in this system, called BUGS (a bug-based search strategy), is due to evolving choice of directions, rather than positions (as with usual real-valued GA methods) [55]. The bugs evolved by the BUGS program learn to search for bacteria in the highest density regions and are thus likely to move toward those points where the function has a maximum value. This strategy combines a hill-climbing mechanism with the adaptive method of GA, and thus overcomes the usual weakness of simple GAs (which do not include such local search mechanisms).

6.6.1 Evolution of predatory behaviors using genetic search

This section introduces the fundamental idea of BUGS. We can experimentally verify the evolution of bugs which possess "predatory behaviors," i.e., the evolution of bugs that learn to hunt bacteria. The original motivation for these experiments was derived from [29]. Bugs learn to move to those regions in the search space where the bacterial concentration is highest. Since the bug concentration is set up to be proportional to the local value of the function to be maximized in the search space, the "stabilized" bug concentrations are proportional to these search space values. Hence the bugs learn (GA style) to be hill climbers. A Swarm-based BUGS simulator is available for readers' self-study. The details are given in Section 6.7.

6.6.1.1 Bugs hunt bacteria

Figure 6.24(a) illustrates the world in which bugs (large dots) live (a 512×512 cellular grid). They feed on bacteria (small dots) which are continually being deposited. The normal bacterial deposition rate is roughly 0.5 bacterium per (GA) generation over the whole grid. Each bug has its internal energy source. The maximum energy supply of a bug is set at 1500 units. When a bug's energy supply is exhausted, the bug dies and disappears. Each bacterium eaten provides a bug with 40 units of energy, which is enough to make 40 moves, where a move is defined to be one of six possible directional displacements of the bug, as shown in Fig. 6.25.

A bug's motion is determined by coded instructions on its gene code. The six directions a bug can move are labeled F, R, HR, RV, HL, and L for Forward, Right, Hard Right, Reverse, Hard Left, and Left, respectively. The GA chromosome format for these bugs is an integer vector of size six where the elements of the vector correspond to the directions in the following order: (F,R,HR,RV,HL,L), e.g., (2,1,1,1,3,2) (as shown in the window in Fig. 6.24(a)). When a bug is to make a move, it will move in the direction d_i (e.g., $d_3 = HR$) with a probability $p(d_i)$, which is determined by the following formula:

$$p(d_i) = \frac{e^{a_i}}{\sum_{j=1}^{6} e^{a_j}} \tag{6.20}$$

where a_i is the ith component value of the chromosome vector (e.g., $a_5 = 3$ above). Once a move is made, a new directional orientation should be determined. Figure 6.25 shows the new F_{next} directions, e.g., if the move is R, the new forward direction will be to the right (i.e., east). For instance, a bug with a gene code of (1,9,1,1,1,1) turns frequently in direction R so that it is highly likely to move in a circle.

After 800 moves (i.e., when it attains an "age" of 800), the bug is said to be "mature" and is ready to reproduce if it is "strong" (i.e., its energy is greater than a threshold value of 1000 energy units). There are two types of reproduction, asexual and sexual (see Fig. 6.26). With asexual reproduction, a strong mature bug disappears and is replaced by two new bugs (in the same

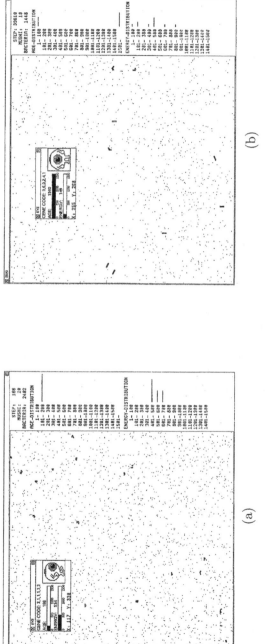

(b)

(a)

FIGURE 6.24: Bug world: (a)166th generation, (b) 39,618th generation.

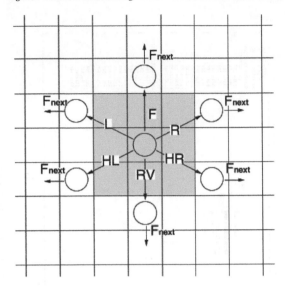

FIGURE 6.25: Bug's gene code.

cell on the grid). Each daughter bug has half the energy of its parent. The genes of each daughter bug are mutated as follows. One of the components of the directional 6-vector is chosen with uniform probability. The value of the direction is replaced by a new value chosen with uniform probability (over the integer range of, e.g., [0,10]). Sexual reproduction occurs when two strong mature bugs "meet" (i.e., they move within a threshold distance from each other called the "reproductive radius"). The distance between two parents is defined as the Euclidean distance between the two parents. The reproductive radius is set at 10.0. The two parents continue to live and are joined by the two daughter bugs. Each parent loses half of its energy in the sexual reproductive process. As a result, two children are born whose energies are half the average of the parents' energies. The children's genes are obtained by applying mutation and uniform crossover operators to the parents' genes. Thus, these reproductions are constrained by probabilities.

Figure 6.24(b) shows the results of the first simple evolutionary experiment. The simulation began with ten bugs with random genetic structures. Most of the bugs jittered from side to side unpredictably and are called "jitterbugs." They are likely to starve to death because they eat up most of the food in their immediate vicinity and are unable to explore widely. In time "cruiser" bugs evolve, which move forward most of the time and turn left or right occasionally. Note that if a bug hits an edge of the grid, it stays there until an appropriate move displaces it away from that grid edge. These "cruiser" bugs succeed in finding food and thus dominate the entire population. A typical chromosome for a "cruiser" bug is shown in the sub-window of Fig. 6.24(b),

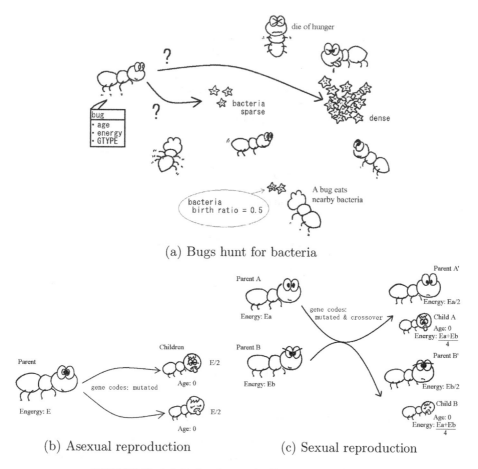

(a) Bugs hunt for bacteria

(b) Asexual reproduction   (c) Sexual reproduction

FIGURE 6.26: A schematic illustration of bugs.

i.e., (9,6,0,2,4,1). The remarkable features of this chromosome vector are as follows:

(1) The forward gene (F) is large (9).

(2) The reverse gene (RV) is small (2).

(3) One of the Right (R), Left (L), Hard Right (HR), and Hard Left (HL) is of moderate size (6).

The second feature is important because bugs with large "reverse" (RV) gene values create "twirlers," which make too many turns in one direction. Such unfortunate creatures usually die. The third feature is also essential, because intelligent bugs have to avoid loitering around wall edges.

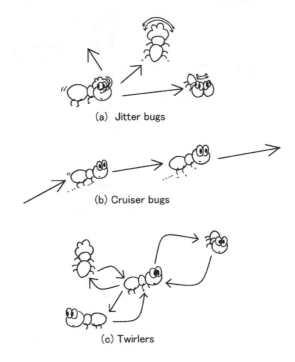

(a) Jitter bugs

(b) Cruiser bugs

(c) Twirlers

FIGURE 6.27: Types of bugs.

6.6.1.2 Effectiveness of sexual reproduction

Dewdney's original paper used only mutation operators, i.e., asexual reproduction. Sexual reproduction is introduced in bugs to increase the effective evolution of bugs [54]. It can be experimentally shown that the speed of evolution is higher with sexual reproduction. In the first experiment, we can statistically compare the performance rates, where the performance of bugs at generation t is defined as follows:

$$Performance(t) = \sum_{i=0}^{9} Perf(t-i), \qquad (6.21)$$

where

$$Perf(k) = \frac{\#Eaten(k)}{\#Bac(k) \times \#Bug(k)} \qquad (6.22)$$

$$\#Bug(k) = \text{(no. of bugs at the kth generation)} \qquad (6.23)$$

$$\#Bac(k) = \text{(no. of bacteria at the kth generation)} \qquad (6.24)$$

$$\#Eaten(k) = \text{(no. of bacteria eaten by bugs at the kth generation)} \qquad (6.25)$$

This indicates how many bacteria are eaten by bugs as a whole in the last

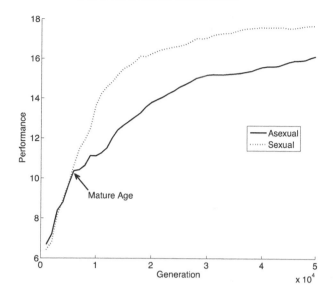

FIGURE 6.28: Performance comparison.

ten generations. As can be seen in Fig. 6.28, sexual reproduction after the mature age (800) performs better than asexual reproduction, which is tested statistically.

In a second experiment, the bacteria in the lower left-hand corner (called the Garden of Eden, a square of 75×75 cells) are replenished at a much higher rate than normal (Fig. 6.29(a)); the normal bacterial deposition rate is roughly 0.5 bacterium per (GA) generation over the whole grid. In the Garden of Eden, this rate is 0.5 over the 75×75 area, i.e., a rate roughly $\frac{512 \times 512}{75 \times 75} = 47$ times greater than normal. As the (GA) generations proceeded, the cruisers evolved as before. But within the Garden of Eden, the jitterbugs were more rewarded for their jittering around small areas (see Fig. 6.27). Thus, two kinds of "species" evolved (i.e., cruisers and twirlers) (Fig. 6.29(c)). Note how typical gene codes of these two species differed from each other. In this second experiment, three different strategies (asexual reproduction, sexual reproduction, and sexual reproduction within a reproductive radius) are compared in four different situations. The aim is to evolve a mix of bugs, namely, the cruisers and twirlers. Two initial conditions are tested: a) randomized initial bugs and b) cruisers already evolved. In addition, the influence of an empty area in which no bacteria exist is investigated. Obviously this empty-area condition makes the problem easier. The results of these experiments are shown in Table 6.4.

As shown in the table, sexual reproduction with a reproductive radius is superior to the other two strategies and the performance improvement is significant for more difficult tasks such as non-empty-area conditions.

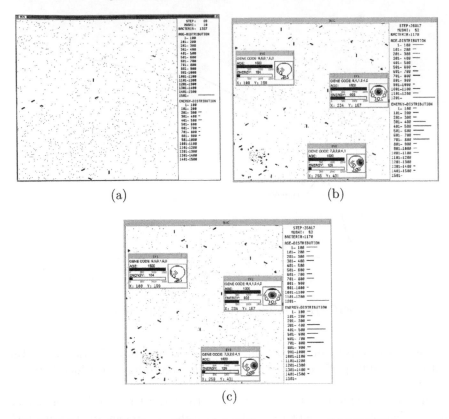

(a)   (b)

(c)

FIGURE 6.29: Garden of Eden (a) 69th generation (b) $72,337$th generation (c) $478,462$nd generation.

Therefore, it was confirmed that crossover is useful for the evolution of predatory behavior. The method described contrasts with traditional GAs in two ways: the bugs' approach uses search directions rather than positions, and selection is based on energy. This idea leads to a bug-based GA search (BUGS) whose implementation is described in the next section.

6.6.2   A bug-based GA search

Those individuals which perform the search in this scheme are called "bugs." The function that these bugs maximize is defined as:

$$f(x_1, x_2, \cdots, x_n) \qquad \text{where } x_i \in Dom_i, \qquad (6.26)$$

where Dom_i represents the domain of the ith parameter x_i.

TABLE 6.4: Experimental results (sexual vs. asexual selection).

Task		Asexual	Sexual	Sexual
initial	empty area	mutation	crossover mutation	Proximity crossover mutation
Random	○	∘	○	⊙
Cruisers	○	∘	○	⊙
Cruisers	×	△	△	○
Random	×	△	△	○

△ difficult   ∘ possible   ○ fast   ⊙ faster

Each bug in the BUGS program is characterized by 3 parameters:

$$Bug_i(t): \text{position} \quad \vec{X}_i(t) = \quad (x_1^i(t), \cdots, x_n^i(t)) \quad (6.27)$$

$$\text{direction} \quad \vec{DX}_i(t) = \quad (dx_1^i(t), \cdots, dx_n^i(t)) \quad (6.28)$$

$$\text{energy} \quad e_i(t) \quad (6.29)$$

where t is the generation count of the bug, x_j is its jth component of the search space, and $\vec{DX}_i$ is the direction in which the bug moves next. The updated position is calculated as follows:

$$\vec{X}_i(t+1) = \vec{X}_i(t) + \vec{DX}_i(t) \quad (6.30)$$

The fitness of each bug is derived with the aid of the function (6.26). The energy $e_i(t)$ of bug i at time (or generation) t is defined to be the cumulative sum of the function values over the previous T time steps or generations, i.e.,

$$e_i(t) = \sum_{k=0}^{T} f(\vec{X}_i(t-k)) \quad (6.31)$$

The format of a bug's "chromosome" which is used in the BUGS program is the bug's real numbered DX vector, i.e., an ordered list of N real numbers. With the above definitions, the BUGS algorithm can now be introduced:

Step 1 The initial bug population is generated with random values:
$Pop(0) = \{Bug_1(0), \cdots, Bug_N(0)\}$
where N is the population size. The generation time is initialized to $t := 1$. The cumulative time period T (called the *Bug-GA Period*) is set to a user-specified value.

Step 2 Move each bug using eq. (6.30) synchronously.

Step 3 The fitness is derived using eq. (6.26) and the energy is accumulated using:
for $i := 1$ to N do $e_i(t) := e_i(t-1) + f_i(\vec{X(t)})$

Step 4 If t is a multiple of T (T-periodical), then execute the GA algorithm (described below) called *BUGS-GA(t)*, and then go to **Step 6**.

Step 5 $Pop(t+1) := Pop(t)$, $t := t+1$, and then go to **Step 2**.

Step 6 for $i := 1$ to N do $e_i(t) := 0$
     $t := t+1$. Go to **Step 2**.

In **Step 1**, initial bugs are generated on conditions that for all i and j,

$$x_j^i(0) := \textbf{Random}(a, b) \tag{6.32}$$

$$dx_j^i(0) := \textbf{Random}(-|a-b|, |a-b|) \tag{6.33}$$

$$e_i(0) := 0 \tag{6.34}$$

where **Random**(a,b) is a uniform random generator between a and b. The *BUGS-GA* period T specifies the frequency of bug reproductions. In general, as this value becomes smaller, the performance becomes better, but at the same time, the convergence time is increased.

For real-valued function optimization, real-valued GAs are preferred over string-based GAs (see [58] for details). Therefore, the real-valued GA approach is used in *BUGS-GA*. The BUGS version of the genetic algorithm, *BUGS-GA(t)*, is as follows (Table 6.5):

Step 1 $n := 1$.

Step 2 Select two parent bugs $Bug_i(t)$ and $Bug_j(t)$ using a probability distribution over the energies of all bugs in $Pop(t)$ so that bugs with higher energy are selected more frequently.

Step 3 With probability P_{cross}, apply the uniform crossover operation to the $\vec{DX}$ of copies of $Bug_i(t)$ and $Bug_j(t)$, forming two offspring $Bug_n(t+1)$ and $Bug_{n+1}(t+1)$ in $Pop(t+1)$. Go to **Step 5**.

Step 4 If **Step 3** is skipped, form two offspring $Bug_n(t+1)$ and $Bug_{n+1}(t+1)$ in $Pop(t+1)$ by making copies of $Bug_i(t)$ and $Bug_j(t)$.

Step 5 With probability P_{asex}, apply the mutation operation to the two offspring $Bug_n(t+1)$ and $Bug_{n+1}(t+1)$, changing each allele in $\vec{DX}$ with probability P_{mut}.

Step 6 $n := n+2$.

Step 7 If $n < N$ then go to **Step 2**.

The aim of the *BUGS-GA* subroutine is to acquire bugs' behavior adaptively. This subroutine works in much the same way as a real-valued GA, except that it operates on the directional vector ($\vec{DX}$), not on the positional

TABLE 6.5: Flow chart of the reproduction process in *BUGS-GA*.

Sexual reproduction:

	individual	energy	gene			individual	energy	gene
Parent$_1$	$Bug_i(t)$	E_1	G_1	$\Rightarrow$	Parent$_1'$	$Bug_n(t+1)$	$(E_1/2)$	G_1
Parent$_2$	$Bug_j(t)$	E_2	G_2	$\Rightarrow$	Parent$_2'$	$Bug_{n+1}(t+1)$	$(E_2/2)$	G_2
					Child$_1$	$Bug_{n+2}(t+1)$	$(E_1 + E_2/4)$	G_1'
					Child$_2$	$Bug_{n+3}(t+1)$	$(E_1 + E_2/4)$	G_2'

Reproduction condition $\Rightarrow$
   $(E_1 >$ Reproduction energy threshold$) \wedge (E_2 >$ Reproduction energy threshold$)$
                $\wedge$ (distance between Parent$_1$ and Parent$_2 <$ Reproduction Radius)

Recombination $\Rightarrow$
   G_1', G_2': uniform crossover of G_1 and G_2 with P_{cross}, mutated with P_{mut}

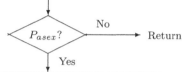

No → Return

P_{asex}?

Yes

Asexual reproduction:

	individual	energy	gene			individual	energy	gene
Parent	$Bug_i(t)$	E	G	$\Rightarrow$	Child$_1$	$Bug_n(t+1)$	$(E/2)$	G'
				$\Rightarrow$	Child$_2$	$Bug_{n+1}(t+1)$	$(E/2)$	G''

Reproduction condition $\Rightarrow (E >$ producible energy$)$

   G', G'': mutation of G with P_{mut}

vector $(\vec{X})$. Positions are thus untouched by the adaptive process of the GA, and are changed gradually as a result of increment by eq. (6.30). On the other hand, fitness is evaluated using positional potential by eq. (6.26), which is the same as for a real-valued GA. Furthermore, chromosome selection is based on cumulative fitness, i.e., energy. The summary of differences between a real-valued GA and BUGS is presented in Table 6.6.

The main difference lies in the GA target (i.e., $\vec{X}$ vs $\vec{DX}$) and the selection criteria (i.e., energy vs. fitness). Remember that the basic idea of this combination is derived from a paper [29] which simulated how bugs learn to hunt bacteria (as described in Section 6.6.1.1).

Figure 6.30 shows the evolution (over 5 to 53 time steps) of the positions and directions of the bugs in the BUGS program. These bugs were used to optimize the following De Jong's $F2$ function (modified to the maximization problem; see Appendix A.2 for details):

$$Maximize \qquad f(x_1, x_2) = -(100(x_1^2 - x_2)^2 + (1 - x_1)^2)$$
$$where \quad -2.047 \leq x_i < 2.048. \qquad (6.35)$$

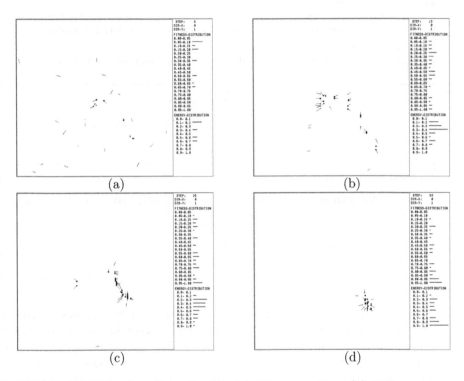

FIGURE 6.30: Bugs' motions for $F2$: (a) 5th generation, (b) 13th generation, (c) 25th generation, (d) 53rd generation.

TABLE 6.6: BUGS vs. a real-valued GAs.

	BUGS	**Real-valued GA**
Fitness evaluation	$\vec{X}$	$\vec{X}$
GA operator target	$D\vec{X}$	$\vec{X}$
Selection criteria	Energy	Fitness

The main window shows a bird's eye view of a two-dimensional projection of the $F2$ function domain. The bugs are represented as black dots. The larger the black dot, the greater the bug's cumulative energy, and hence its fitness. The "tail" of a bug indicates the direction vector of its motion. The energy and fitness distributions are shown as bar graphs to the right of the windows. It is clear that each bug climbs the ($F2$) potential hill. As the generations pass (i.e., the time steps), the tails become shorter and shorter (as shown in Fig. 6.30(b),(c),(d)). This shows that the bugs evolved the correct directions in which to move, and thus "converged" to the top of the hill. It was confirmed that for optimizing various benchmark functions, the approach used in the BUGS program has the same or greater search power than real-valued GAs [54, 55, 58].

As the generations proceed, the bugs in the BUGS program are converged to the top of the hill via position-based fitness selection (see Fig. 6.31). In the meantime, the motion directions are gradually refined by the GA mechanism. We consider this BUGS-type adaptation to be closely related to the building-block hypothesis and to a schema-based adaptation. In some cases, the BUGS approach has been shown to be a more efficient search strategy than a real-valued GA. This is justified by the comments of Rechenberg [100]. He claimed that, in general, the essential dimensions for search are relatively few in higher dimensions and that an effective optimization is realized by following the gradient curve in these essential dimensions. We think that direction-based (rather than position-based) building blocks produce a more efficient search strategy in the sense that essential dimensions are adaptively acquired in bugs' motions, similar to a PSO search.

6.7 BUGS in Swarm

This program simulates the evolution of predatory activity of bugs. Figure 6.32 is a screenshot of the program.

The bugs are colored green (young) or yellow (adults). They survive by consuming bacteria (the red dots), which are randomly generated throughout the space. A bug obtains 40 units of energy for each bacterium it consumes. A bug uses up one unit of energy for each step it moves, and dies if its energy reserve reaches zero. It must optimize its method of motion in order to live a long life. Its motion is determined by genetic codes; these consist of the six integer values expressing the directions {Forward, Right, Hard Right, Reverse, Hard Left, Left} (see Fig. 6.25). The greater the value of any one integer, the higher the probability that the individual will move in that direction. An individual reaches maturity after 800 steps, and after its energy exceeds 1,000, it engages in sexual reproduction. In this process, if there is another equally mature and strong bug nearby, the two individuals perform genetic crossover and mutation, and produce two child bugs. The energy of each parent is then cut in half and each child receives one-half the mean energies of the parents. If an individual has traveled 1,000 steps and has over 1,300 units of energy, it engages in asexual reproduction. In asexual reproduction, the parent's genetic code is mutated to produce two child bugs, and the parent dies. Each child receives one-half the parent's energy.

The lower right region of the figure is called the "Garden of Eden"; it has a higher generation rate of bacteria. The bugs near the Garden of Eden evolve a tendency to move so as to remain within this area. The "Eden" parameter in the "ModelSwarm" probe activates or deactivates the Garden of Eden.

The age of bugs in a population (how many steps they have moved) and the distribution of energy are shown on a bar graph. The horizontal axis represents age (energy) and the vertical axis represents the number of bugs.

The following buttons are used to operate BUGS:

- Start button: Initiates the simulation.

- Stop button: Temporarily stops the simulation.

- Next button: Advances the stimulation to the next time step.

- Save button: Cannot be used in this simulation.

- Quit button: Stops the simulation.

The following relationships are included in the "ModelSwarm" parameter probes:

- worldXSize: Extent of space (horizontal axis)

- worldYSize: Extent of space (vertical axis)

- seedProb: Growth rate of bacteria under initial conditions

- bugDensity: Growth rate of bugs under initial conditions

- Eden: Application of the Garden of Eden, i.e., 1 (Yes) or 0 (No)

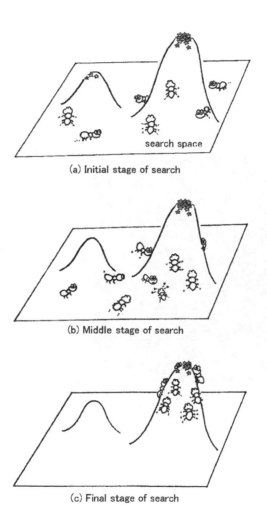

(a) Initial stage of search

(b) Middle stage of search

(c) Final stage of search

FIGURE 6.31: Illustration of bug-based search.

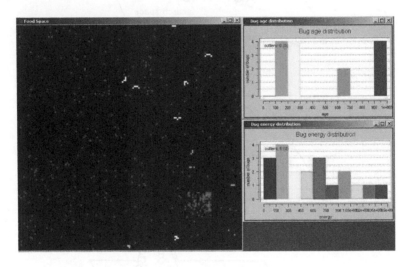

FIGURE 6.32 (See Color Insert): BUGS simulator.

Chapter 7

Cellular Automata Simulation

This intelligent behavior would be just another one of those organizational phenomena like DNA which contrived to increase the probability of survival of some entity. So one tends to suspect, if oneȦfs not a creationist, that very very large LIFE configurations would eventually exhibit intelligent [characteristics]. Speculating what these things could know or could find out is very intriguing ... and perhaps has implications for our own existence [79, pp.139–140].

7.1   Game of life

The eminent mathematician John von Neumann studied self-reproducing automata in 1946, shortly before his death. He found that self-reproduction is possible with 29 cell states, and proved that a machine could not only reproduce itself, but could also build machines more complex than itself. The research stopped because of his death; however, in 1966, Arthur Burks edited and published von Neumann's manuscripts. John Conway, a British mathematician, expanded on the work of von Neumann and, in 1970, introduced the Game of Life, which attracted immense interest. Some people became "Game of Life hackers," programmers and designers more interested in operating computers than in eating; they were not the criminal hackers of today. Hackers at MIT rigorously researched the Game of Life, and their results contributed to advances in computer science and artificial intelligence. The concept of the Game of Life evolved into the "cellular automaton" (CA), which is still widely studied in the field of artificial life. Most of the research on artificial life shares much in common with the world where hackers played in the early days of computers.

The Game of Life is played on a grid of equal-sized squares (cells). Each cell can be either "on" or "off." There are eight adjacent cells to each cell in a two-dimensional grid (above and below, left and right, four diagonals). This is called the Moore neighborhood. The state in the next step is determined by the rules outlined in Table 7.1. The "on" state corresponds to a "•" in the cell, whereas the "off" state corresponds to a blank. The following interesting patterns can be observed with these rules.

TABLE 7.1: State of cell in the next step.

Current state of cell	States of neighbor cells	State in the next step
On	two or three are "on"	On
	Other cases	Off
Off	three are "on"	Off
	Other cases	Off

(1) Disappearing pattern (diagonal triplet)

$$\begin{matrix} \bullet & & & & \\ & \bullet & \Rightarrow & \bullet & \Rightarrow & \text{disappears} \\ & & \bullet & & \end{matrix} \qquad (7.1)$$

(2) Stable pattern (2 × 2 block)

$$\begin{matrix} \bullet\bullet \\ \bullet\bullet \end{matrix} \Rightarrow \begin{matrix} \bullet\bullet \\ \bullet\bullet \end{matrix} \Rightarrow \begin{matrix} \bullet\bullet \\ \bullet\bullet \end{matrix} \Rightarrow \text{remains stable} \qquad (7.2)$$

(3) Two-state switch (Flicker, where vertical triplets and horizontal triplets appear in turn)

$$\begin{matrix} \bullet \\ \bullet \\ \bullet \end{matrix} \Rightarrow \bullet\bullet\bullet \Rightarrow \begin{matrix} \bullet \\ \bullet \\ \bullet \end{matrix} \Rightarrow \text{repeats} \qquad (7.3)$$

(4) Glider (pattern moving in one direction)

$$\Rightarrow \quad \Rightarrow \quad \Rightarrow \quad \text{moves to bottom right} \qquad (7.4)$$

"Eaters" that stop gliders and "glider guns" that shoot gliders can be defined, and glider guns are generated by the collision of gliders. Such self-organizing capabilities mean that the Game of Life can be used to configure a universal Turing machine. The fundamental logic gates (AND, OR, NOT) consist of glider rows and disappearing reactions, and blocks of stable patterns are used as memory. However, the number of cells needed in a self-organizing system is estimated at about 10 trillion (3 million × 3 million). The size would be a square whose sides are 3 km long, if 1 mm^2 cells are used.

Consider a one-dimensional Game of Life, one of the simplest cellular automata. The sequence of cells in one dimension at time t is expressed as follows:

$$a_t^1, a_t^2, a_t^3, \cdots \tag{7.5}$$

Here, each variable is either 0 (off) or 1 (on). The general rule used to determine the state a_{t+1}^i of cell i at time $t+1$ can be written as a function F of the state at time t as

$$a_{t+1}^i = F(a_t^{i-r}, a_t^{i-r+1}, \cdots, a_t^i, \cdots, a_t^{i+r-1}, a_t^{i+r}) \tag{7.6}$$

Here, r is the radius, or range of cells that affects this cell.

For instance, a rule for $r = 1$,

$$a_{t+1}^i = a_t^{i-1} + a_t^i + a_t^{i+1} \pmod 2 \tag{7.7}$$

results in the determination of the next state as follows:

```
time t    :    0010011010101100
time t+1  :    *11111001010001*
```

An interesting problem is the task of finding the majority rule. The task is to find a rule that would ultimately end in a sequence of all 1 (0) if the majority of the cells are 1 (0) with the minimum radius (r) possible for a one-dimensional binary sequence of a given length. The general solution to this problem is not known.

A famous example is a majority rule problem with length 149 and radius 3. The problem is reduced to finding a function that assigns 1 or 0 to an input with 7 bits ($= 3 + 1 + 3$, radius of 3 plus itself); therefore, the function space is 2^{2^7}.

How can a cellular automaton (CA) obtain a solution to the majority problem?

One method is to change the color (black or white) of a cell to the majority of its neighboring cells. However, this method does not work well, as shown in Fig.7.1 because it results in a fixed pattern divided into black and white.

In 1978, Gacs, Kurdyumov [37A], and Levin found the rules (GLK) regarding this problem. Lawrence Davis obtained an improved version of these rules in 1995, and Rajarshi Das proposed another modification. There is also research to find effective rules through GA or GP. The concept of Boolean functions is applied when GP is used. The fitness value is defined by the percentage of correctly processed sequences out of 1000 randomly generated sequences of length 149.

Rules determined by various methods are summarized in Table 7.2. Here, the transition rules are shown from 0000000 to 1111111 in 128-bit form. In other words, if the first bit is 0,

$$F(000\ 0\ 000) = 0 \tag{7.8}$$

Table 7.3 is a comparison of different rules. The rules obtained using GP were very effective; reference [2] contains the details of this work.

FIGURE 7.1 (See Color Insert): CA carrying out majority voting (with permission of Oxford University Press, Inc. [88]).

FIGURE 7.2: Behavior of CA driven by GA (with permission of Oxford University Press, Inc. [88]).

Fig.7.2 shows how a CA obtained by GA can solve this problem well [87, 88]. The regions that were initially dominated by black or white cells become regions that are completely occupied by either black or white cells. A vertical line always exists at locations where a black region to the right meets a white region to the right. In contrast, a triangular region with a chessboard pattern forms where a white region to the right meets a black region to the right.

The two edges of the growing triangular region at the center with a chessboard pattern grow at the same pace, progressing the same distance per unit time. The left edge extends until it collides with a vertical boundary. The right edge barely avoids the vertical boundary at the left (note that the right and left edges are connected). Therefore, the left edge can extend for a shorter length, which means that the length of the white region limited by the left edge is shorter than the length of the black region limited by the right edge. The left edge disappears at the collision point, allowing the black region to grow.

TABLE 7.2: Majority rules.

Name of rule (year)	Transition rules
GKL (1978)	00000000 01011111 00000000 01011111 00000000 01011111 00000000 01011111 00000000 01011111 11111111 01011111 00000000 01011111 11111111 01011111
Davis (1995)	00000000 00101111 00000011 01011111 00000000 00011111 11001111 00011111 00000000 00101111 11111100 01011111 00000000 00011111 11111111 00011111
Das (1995)	00000111 00000000 00000111 11111111 00001111 00000000 00001111 11111111 00001111 00000000 00000111 11111111 00001111 00110001 00001111 11111111
GP (1995)	00000101 00000000 01010101 00000101 00000101 00000000 01010101 00000101 01010101 11111111 01010101 11111111 01010101 11111111 01010101 11111111

TABLE 7.3: Performance in the majority problem.

Rule	Performance	Number of tests
GKL	81.6%	10^6
Davis	81.8%	10^6
Das	82.178%	10^7
GA	76.9%	10^6
GP	82.326%	10^7

Furthermore, the two edges disappear at the bottom vertex and the entire lattice row becomes black, showing that the correct answer was obtained.

Melanie Mitchell analyzed the information processing structure on CA that evolved through GA by using the behavior of dynamic systems [87, 88]. The boundaries between simple regions (edges and vertical boundaries) are considered carriers of information, and information is processed when these boundaries collide. Figure 7.3 shows only the boundaries in Fig. 7.2. These boundary lines are called "particles" (similar to elementary particles in a cloud chamber used in physics). The particles are represented by Greek letters following the tradition in physics. Six particles are generated in this CA. Each particle represents a different type of boundary. For instance, η is the boundary between a black region and a chessboard-patterned region. A number of collisions of particles can be observed. For example, $\beta + \gamma$ results in the generation of a new particle η, and both particles annihilate in $\mu + \eta$.

It is easy to understand how information is coded and calculated when the behavior of CA is expressed in the language of particles. For instance, α and β particles are coded with different information on the initial configuration.

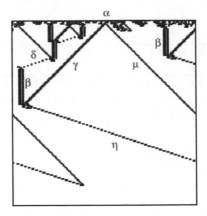

FIGURE 7.3: Explanation of CA behavior from collision of particles (with permission of Oxford University Press, Inc. [88]).

γ particles contain information that this is the boundary with a white region, and a μ particle is a boundary with a white region. When a γ particle collides with a β particle before colliding with a μ particle, this means that the information carried by the β and γ particles becomes integrated, showing that the initial large white region is smaller than the initial large black region that shares a boundary. This is coded into the newly generated η particle.

Stephen Wolfram [128] systematically studied the patterns that form when different rules (eq. (7.6)) are used. He grouped the patterns generated by one-dimensional CA into four classes.

Class I All cells become the same state and the initial patterns disappear. For example, all cells become black or all cells become white.

Class II The patterns converge into a striped pattern that does not change or a pattern that periodically repeats.

Class III Aperiodic, chaotic patterns appear.

Class IV Complex behavior is observed, such as disappearing patterns or aperiodic and periodic patterns.

Examples of these patterns are shown in Fig. 7.4. The following are the rules behind these patterns (radius 1):

- Class I: Rule 0

- Class II: Rule 245

- Class III: Rule 90

- Class IV: Rule 110

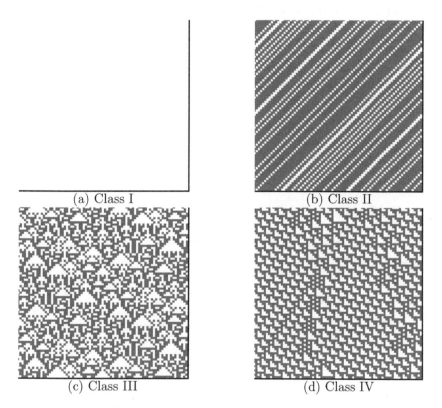

(a) Class I   (b) Class II

(c) Class III   (d) Class IV

FIGURE 7.4: Examples of patterns.

TABLE 7.4: Rule 184.

$a_t^{i-1}, a_t^i, a_t^{i+1}$	111	110	101	100	011	010	001	000
a_{t+1}^i	1	0	1	1	1	0	0	0

Here, transition rules are expressed in 2^{2^3} bits from 000 to 111, and the number of the rule is the decimal equivalent of those bits. For instance, the transition rules for Rule 110 in Class IV are as follows [83]:

$$01101110_{binary} = 2 + 2^2 + 2^3 + 2^5 + 2^6 = 110_{decimal} \qquad (7.9)$$

In other words, 000, 100, and 111 become 0, and 001, 010, 011, 101, and 110 become 1.

Rule 110 has the following interesting characteristics in computer science.

(1) Is computationally universal [128].

(2) Shows $1/f$ fluctuation [91].

(3) Prediction is P-complete [90].

Kauffman proposed the concept of the "edge of chaos" from the behavior of CA as discussed above [67]. This concept represents Class IV patterns where periodic patterns and aperiodic, chaotic patterns are repeated. The working hypothesis in artificial life is "life on the edge of chaos."

7.1.1 Rule 184

Rule 184 is known as the Burgers cell automaton (BCA), and has the following important characteristics (Table 7.4).

(1) The number of 1s is conserved.

(2) Complex changes in the 0-1 patterns are initially observed; however, a pattern always converges to a right-shifting or left-shifting pattern.

Take a_t^i as the number of cars on lattice point i at time t. The number is 1 (car exists) or 0 (no car exists). The car on lattice point i moves to the right when the lattice point is unoccupied, and stays on the lattice point when occupied at the next time step. This interpretation allows a simple description of traffic congestion using rule 184.

BCA can be expanded as follows:

- BCA with signals: restrict traffic to the right at each lattice point

- Probabilistic BCA: move with a probability

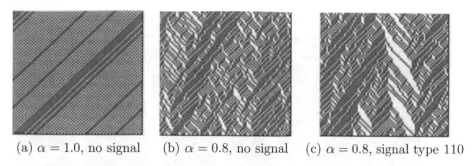

(a) $\alpha = 1.0$, no signal   (b) $\alpha = 0.8$, no signal   (c) $\alpha = 0.8$, signal type 110

FIGURE 7.5: Congestion simulation using BCA.

An in-depth explanation of CAs describing traffic congestion is given in section 7.9.

Figure 7.5 shows a result of BCA simulation. Cars are shown in black in this simulator. Continuous black areas represent congestion because cars cannot move forward. Congestion is more likely to form when the probability that a car moves forward α is small. Figure 7.5(a) shows how congestion forms (black bands wider than two lattice points form where a car leaves the front of the congestion and another car joins at the rear). A signal is added at the center in Figure 7.5(c). The signal pattern is shown in blue when 1 and red when 0, and the pattern in this simulation was 110 (i.e., blue, blue, red, blue, blue, red,...).

The Swarm simulator using BCA is an application of one-dimensional CA, and the parameter `probe` in `ObserverSwarm` displays the following parameters.

- `worldSizeX`: size of the space (horizontal axis)

- `worldSizeY`: size of the space (vertical axis)

- `Alpha`: probability that a car moves to the right

- `Signal1`: pattern of signal 1

- `Signal2`: pattern of signal 2

The probability that cars move to the right and the pattern of signals can be changed during runs by clicking the "Stop" button, changing values, and re-running. Here, after changing the parameters, you need to press "Enter" and click the `applyRule` button.

A simulation of silicon traffic based on these models is shown in Fig. 7.6. A two-dimensional map is randomly generated, and cars are also placed at random. Two points are randomly chosen from the nodes on the map and are designated as the origin and the destination. The path from the origin to the destination is determined by Dijkstra's algorithm. Here, a cost calculation where the distances between nodes are weighted with the density of cars in each edge is performed to determine a path that avoids congestion. The signal

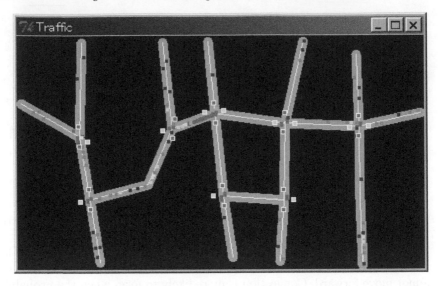

FIGURE 7.6 (See Color Insert): Simulation of silicon traffic.

patterns are designed to change the passable edge in a given order at a constant time interval.

7.2　Conway class with Swarm

A standard class to describe cellular automaton is

`Class ConwayLife2dImpl`

The advantages of using the `Conway` class are

(1) The description of a cell automaton is easy.

(2) Faster execution of update rules is possible.

For example, let us make a life game program (Fig. 7.7). You might guess that in this game cells are spread all over the square lattice, and either take up life (1, on) or death (0, off). The state of cells is updated by the following rules:

(1) If the state of cells is death, it comes back to life next time if out of the eight surrounding places three are alive.

(2) If the state of cells is life, it stays alive next time if out of the eight surrounding places two or three are alive. Else, it dies in the next time.

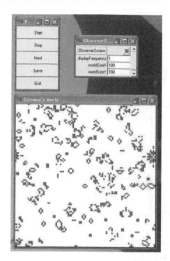

FIGURE 7.7: Game of life.

As before, implementation by allocating one `Bug` (object) to one cell can also be done. In this case, update rules are described in the "`step`" method of `Bug.java` as follows. This can be thought of as an application of `simpleObserverBug2`.

```
public void step(){
    int i,j;
    int sx, sy;
    int num=0;
    Bug b;
// Number of live cells around is obtained in ''sum''
  for(i=xPos-1;i<xPos+2;i++){
        for(j=yPos-1;j<yPos+2;j++){
            if( !(i==xPos && j==yPos) ){
                sx = (i+worldXSize)%worldXSize;
                sy = (j+worldYSize)%worldYSize;
                b = (Bug) world.getObjectAtX$Y(sx, sy);
                if( b.isAlive() ) num++;
            }
        }
    }
// Execution of update rules of life and death
    if(is_alive){
        if(num==2 || num==3){
            next_is_alive=true;
        }else{
            next_is_alive=false;
```

```
        }
    }else{
        if(num==3){
            next_is_alive = true;
        }else{
            next_is_alive = false;
        }
    }
}
```

On the other hand, writing lattice plane and rules in one class using "ConwayLife2dImpl" is also possible. Let us look at the stepRule method inside ConwayWorld.java. Here, all coordinates (x, y),

```
sum += this.getValueAtX$Y(xm1, ym1);   // down left
sum += this.getValueAtX$Y(x,ym1);     // down
sum += this.getValueAtX$Y(xp1,ym1);   // lower right
sum += this.getValueAtX$Y(xm1,y);     // left
sum += this.getValueAtX$Y(xp1, y);    // right
sum += this.getValueAtX$Y(xm1, yp1);  // upper left
sum += this.getValueAtX$Y(x, yp1);    // up
sum += this.getValueAtX$Y(xp1, yp1);  // upper right
```

count the number of lives (1) in cells in the eight neighbors. However,

```
xm1 = (x + xsize - 1) % xsize; // left
xp1 = (x + 1) % xsize;         // right
ym1 = (y + ysize - 1) % ysize; // down
yp1 = (y + 1) % ysize;         // up
```

Dividing by xsize, ysize and obtaining the remainder is because it is considered to be connecting the vertical and horizontal (torus structure) of the lattice plane. Moreover, the part below calculates the next state:

```
if(this.getValueAtX$Y(x,y)==1) // am I alive (1) ?
    newState = (sum==2 || sum==3) ? 1 : 0;
else // if I am dead (0)
    newState = (sum==3) ? 1 :0;
```

Let us try to implement the cellular automaton explained in Section 7.1 by applying Bug (Fig. 7.8). This sample program is a 2 state 3 neighbor one-dimensional automaton, and update rules are according to odd parity. In other words, it takes the majority state in the three cells as the next state. In fact, a process such as the one given below is performed by laying Bug (object) in the last line of a two-dimensional FoodSpace grid:

Step1 Bug looks at the state of the bait placed in its neighborhood.

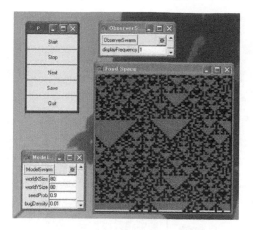

FIGURE 7.8 (See Color Insert): One-dimensional cellular automaton.

Step2 This state decides whether the bait should be placed in the next time or not.

Step3 FoodSpace is shifted one space up.

Step4 Bug places the food.

Step5 Return to **Step1**.

Note that to change the rules of state transition, the contents of the method "step" of Bug.java, which decides the next state, should be changed. For example, it is as follows in the sample program:

```java
public void step(){
    int x;
    int sx;
    int[ ] v = new int [3];
// radius r is 1, therefore, an array of r + 1 + r =3 is prepared
    // xPos is its own coordinates
    // the three values xPos-1, xPos, xPos+1 are substituted in
        v[0], v[1], v[2]
    for(x = xPos-1; x < xPos+2; x++){
        sx = (x + worldXSize) % worldXSize;
        v[x - (xPos-1)] = foodSpace.getValueAtX$Y(sx, yPos);
    }
    // nextvalue is the next state of xPos
// next state is obtained after checking the parity of v[0], v[1],
    v[2]
    if( (v[0]+v[1]+v[2])%2 ==1 )
        nextvalue = 1;
```

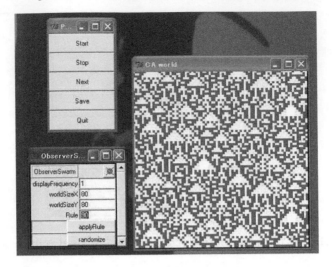

FIGURE 7.9: Execution of Wolfram's experiment.

```
    else
        nextvalue = 0;
}
```

Furthermore, in order to increase the states, follow the steps below:

(1) Add the number of colors equal to the number of states you want to use in `colorMap` inside `ObserverSwarm.java`.

(2) Increase the types of bait equal to the number of states you want to use inside `Bug.java`.

(3) Modify the state transition rules accordingly.

To do the experiment by Wolfram as explained in Section 7.1, let us try to improve the program. The execution example of the system for that is shown in Fig. 7.9. Since in this program, the rule number can be entered in `"Rule,"` observation of the behavior of various classes mentioned above is also possible. Rule number can also be changed during execution, by the following steps:

(1) If during execution, press the Stop button.

(2) Enter the number to be changed in the `"Rule"` field. Then press the Enter key.

(3) Press the `"applyRule"` button.

(4) If required when you press the `"randomize"` button, the cell state is initialized randomly.

(5) Execution starts with new conditions when the "Start" button is pressed.

As the last example, we will explain the self-replicating loop of Langton. This is a two-dimensional cell automaton with eight states and five neighbors, and it is something that replicates its own pattern through 219 transition rules (refer to Table 7.5). In this chart, the sequence of six numbers represents one transition rule (in order, state before transition, states of four adjacent cells, and state after transition). Four cells' states are considered to be clockwise, and may begin in any direction. For example, in the last rule: 7 0272 0 shows that the next state of the center cell, i.e., 7, will be 0 for the following four cases:

0	2	7	2
272	770	272	077
7	2	0	2

Please note that, in the infinite space, situations not in transition states do not occur. In fact, since the space is a torus structure, the occurrence of situations not in the transition rules is possible. However, in that case the state is updated to 0.

As defined in `"initvalue"` in `FoodSpace.java`, the initial state pattern is a loop as follows:

```
2 2 2 2 2 2 2 2
2 1 7 0 1 4 0 1 4 2
2 0 2 2 2 2 2 2 0 2
2 7 2           2 1 2
2 1 2           2 1 2
2 0 2           2 1 2
2 7 2           2 1 2
2 1 2 2 2 2 2 2 1 2 2 2 2 2
2 0 7 1 0 7 1 0 7 1 1 1 1 1 2
  2 2 2 2 2 2 2 2 2 2 2 2 2
```

Here, the part without numbers is the part with state 0. How self-replication goes on through transition of initial patterns is shown in Fig. 7.10. The repetition of the extended protruded part bending to the left and the continuation of the self-replication of the initial state loop can be observed. Cells' states loosely mean the following:

- State 0: Background with nothing

- State 1: Transmission passage of signal

- State 2: Transmission passage's covered portion

- State 4 to 7: Transmitted signal

TABLE 7.5: Self-reproducing loop transition rules

0	0000	0	0	0001	2	0	0002	0	0	0003	0	0	0005	0
0	0006	3	0	0007	1	0	0011	2	0	0012	2	0	0013	2
0	0021	2	0	0022	0	0	0023	0	0	0026	2	0	0027	2
0	0032	0	0	0052	5	0	0062	2	0	0072	2	0	0102	2
0	0112	0	0	0202	0	0	0203	0	0	0205	0	0	0212	5
0	0222	0	0	0232	2	0	0522	2	0	1232	1	0	1242	1
0	1252	5	0	1262	1	0	1272	1	0	1275	1	0	1422	1
0	1432	1	0	1442	1	0	1472	1	0	1625	1	0	1722	1
0	1725	5	0	1752	1	0	1762	1	0	1772	1	0	2527	1
1	0001	1	1	0006	1	1	0007	7	1	0011	1	1	0012	1
1	0021	1	1	0024	4	1	0027	7	1	0051	1	1	0101	1
1	0111	1	1	0124	4	1	0127	7	1	0202	6	1	0212	1
1	0221	1	1	0224	4	1	0226	3	1	0227	7	1	0232	7
1	0242	4	1	0262	6	1	0264	4	1	0267	7	1	0271	0
1	0272	7	1	0542	7	1	1112	1	1	1122	1	1	1124	4
1	1125	1	1	1126	1	1	1127	7	1	1152	2	1	1212	1
1	1222	1	1	1224	4	1	1225	1	1	1227	7	1	1232	1
1	1242	4	1	1262	1	1	1272	7	1	1322	1	1	2224	4
1	2227	7	1	2243	4	1	2254	7	1	2324	4	1	2327	7
1	2425	5	1	2426	7	1	2527	5	2	0001	2	2	0002	2
2	0004	2	2	0007	1	2	0012	2	2	0015	2	2	0021	2
2	0022	2	2	0023	2	2	0024	2	2	0025	0	2	0026	2
2	0027	2	2	0032	6	2	0042	3	2	0051	7	2	0052	2
2	0057	5	2	0072	2	2	0102	2	2	0112	2	2	0122	2
2	0142	2	2	0172	2	2	0202	2	2	0203	2	2	0205	2
2	0207	3	2	0212	2	2	0215	2	2	0221	2	2	0222	2
2	0227	2	2	0232	1	2	0242	2	2	0245	2	2	0252	0
2	0255	2	2	0262	2	2	0272	2	2	0312	2	2	0321	6
2	0322	6	2	0342	2	2	0422	2	2	0512	2	2	0521	2
2	0522	2	2	0552	1	2	0572	5	2	0622	2	2	0672	2
2	0712	2	2	0722	2	2	0742	2	2	0772	2	2	1122	2
2	1126	1	2	1222	2	2	1224	2	2	1226	2	2	1227	2
2	1422	2	2	1522	2	2	1622	2	2	1722	2	2	2227	2
2	2244	2	2	2246	2	2	2276	2	2	2277	2	3	0001	3
3	0002	2	3	0004	1	3	0007	6	3	0012	3	3	0042	1
3	0062	2	3	0102	1	3	0122	0	3	0251	1	4	0112	0
4	0122	0	4	0125	0	4	0212	0	4	0222	1	4	0232	6
4	0252	0	4	0322	1	5	0002	2	5	0021	5	5	0022	5
5	0023	2	5	0027	2	5	0052	0	5	0202	2	5	0212	2
5	0215	2	5	0222	0	5	0224	4	5	0272	2	5	1212	2
5	1222	0	5	1242	2	5	1272	2	6	0001	1	6	0002	1
6	0212	0	6	1212	5	6	1213	1	6	1222	5	7	0007	7
7	0112	0	7	0122	0	7	0125	0	7	0212	0	7	0222	1
7	0225	1	7	0232	1	7	0252	5	7	0272	0			

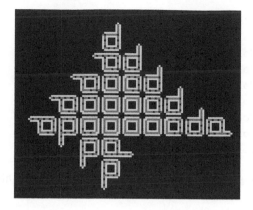

FIGURE 7.10: Self-replicating loop of Langton.

The Swarm implementation of the Langton loop on the two-dimensional FoodSpace grid contains eight types of bait, while only bug is updated. However, state [ ] stores (5 digit number) the current state (own state and 4 neighbors' states). The array that represents the next state is nextstate[ ]. Here, if the current state is represented by x (5 digit number),

```
state[i]==x
```

finds i, and the next state x_next is decided as follows:

```
x_next = nextstate[i]
```

This process is executed by the method getRuleArrayValue, as shown below, inside Bug.java.

```java
private int getRuleArrayValue(int v[]){
    int val=-1;
    int i;
    int index=getMinIndex(v);
    index+=v[0]*10000;
    for(i=0;i<state.length;i++){
        if(state[i]==index){
            val = nextstate[i];
            break;
        }
    }
    return val;
}
```

However, getMinIndex is a process to implement the transition rule shown in Table 7.5 and it is alright if it matches any of the 4 directions, in a clockwise direction. The colors of 8 states are defined as below in buildObjects of ObserverSwarm.java.

```
colorMap=new ColormapImpl(this);
colorMap.setColor$ToName((byte)0,"black");
colorMap.setColor$ToName((byte)1,"red");
colorMap.setColor$ToName((byte)2,"green");
colorMap.setColor$ToName((byte)3,"red");
colorMap.setColor$ToName((byte)4,"red");
colorMap.setColor$ToName((byte)5,"red");
colorMap.setColor$ToName((byte)6,"red");
colorMap.setColor$ToName((byte)7,"red");
```

In other words, state 0 is black, state 2 is green, and all others are being turned red.

7.3   Program that replicates itself

Making a program that outputs its source code looks easy but is a very difficult problem. In particular, care is necessary in treating newline characters and spaces. The following is an example of a self-replicating code in C.

```
#include <stdio.h>
char *s="#include <stdio.h>%cchar *s=%c%s%c;main(){printf(s,10,34,s,34,10);return 0;}%c";main(){printf(s,10,34,s,34,10);return 0;}
```

Character strings are used effectively. There should be no carelessly inserted newline characters in this code: when a newline character is inserted, the code must be changed to output this additional newline character. Note that 10 and 34 in ASCII code are the newline and " characters, respectively. Therefore, executing `printf(s,10,34,s,34,10);` in the `main` function prints the first and second lines of the code.

To confirm that the above is a self-replicating code, copy the above file as a source file such as `ev.c` and execute the following. You can confirm that the output file and the source code are the same (it is recommended that you use the `diff` command that displays the difference between two files). The following is an example of how the self-replicating code is executed.

```
iba@fs(~/tmp)[514]: cat ev.c
#include <stdio.h>
char *s="#include <stdio.h>%cchar *s=%c%s%c;main(){printf(s,10,34,s,34,10);return 0;}%c";main(){printf(s,10,34,s,34,10);return 0;}
iba@fs(~/tmp)[515]: gcc ev.c
iba@fs(~/tmp)[516]: ./a.out > ev2.c
iba@fs(~/tmp)[517]: diff ev.c ev2.c
```

It is also possible to compile code that outputs its source code in languages other than C. For example, the following line in Lisp which is used in artificial intelligence,

```
(setf f '(lambda (x) `(,x ',x)))
```

results in self-replication. The famous computer scientist Knuth said the following in his lecture after receiving the Turing award, the equivalent of the Nobel Prize in computer science [74, p. 672]:

> Some years ago the students at Stanford were excited about finding the shortest FORTRAN program which prints itself out, in the sense that the program's output is identical to its own source text. The same problem was considered for many other languages. I don't think it was a waste of time for them to work on this.

7.4 Simulating forest fires with Swarm

Simulations based on CA are applied in various fields such as the following:

- Life sciences, especially in heredity, immunity, ecological formation

- Prediction of freeway congestion (silicon traffic)

- Disasters: marine pollution from oil spills, forest fires

- Materials and manufacturing

- Fractal formation

These simulations are considered to be an effective method for observing critical behavior in phase transitions.

This section is on the simulation of disasters. The area over which a forest fire can spread depends on the percentage of empty space with no trees. A CA model can therefore be used to observe the ratio between the percentage of empty space and the affected area.

The forest fire model uses CA and follows these rules:

(1) There are three states: trees, fire, and empty space.

(2) Trees can grow on empty space with a fixed probability.

(3) Trees catch fire with a fixed probability.

(4) Fire extends to neighboring trees.

(5) There will be no fire on an empty space.

(6) Spaces will be empty after the fire.

This simple model can show how a fire will spread.

The following is a specific algorithm used in a forest fire model. Cells on a square lattice can be either trees (value 1), trees on fire (value 2), or empty space (value 0). All cells are updated under the following rules:

- If the cell has trees:
  - It will catch fire with probability $(1 - g)$ if at least one adjacent cell is on fire.
  - It will catch fire with probability $f \times (1 - g)$ if no adjacent cells are on fire.
- A cell on fire will become an empty space.
- Empty space will become trees with probability p.

Here, the parameters are the probability of catching fire f, immunity g, and the probability of tree growth p. The following parameter values are used: probability of catching fire $f = 0$, immunity $g = 0.01$, and probability of tree growth $p = 0$. Figure 7.11 shows the area of forest fire with respect to various percentages F of area with trees in the forest before the fire starts. Red cells are fire, green cells are trees, and black cells are empty space in the figure. The large black area indicates the burned area.

The experiments showed that the fire spread over the entire forest when $F > 0.59$ and did not when $F < 0.59$. Threshold values such as $F = 0.59$ are called critical values and are important parameters in the simulation of complex systems.

7.4.1 Simulation of forest fires

A forest fire model is simulated using Swarm. Here, the intermediate state is newly defined by extending the model of the three traditional states (trees/wood, fire, nothing). In other words, a total of seven states—three states of fire and forest each and a state of nothing—are taken to simulate a more complex model.

The definition of fire and states, and the list of parameters used are shown in Tables 7.6 and 7.7, respectively. If we assume our own cell state of "state" of four neighboring cells as N1, N2, N3, N4 and the random number in the range [0,1] is "rand," then the rules of fire will be as follows:

```
if (state=0 & rand<grow1) then state=1
if (state=1 & rand<grow2) then state=2
if (state=2 & rand<grow3) then state=3
if (state=4 & rand<cool1) then state=5
if (state=5 & rand<cool2) then state=6
if (state=6 & rand<cool3) then state=0
if (state=3 & rand<fire)  then state=4
if (state=3 &
        (N1=4 or N2=4 or N3=4 or N4=4) & rand<diffuse)
                          then state=4
```

(a)   (b)

FIGURE 7.11 (See Color Insert): Forest fire examples.

TABLE 7.6: State of forest fire.

State	Display color	Meaning	Explanation
0	black	Ash	State of nothing
1	light green	Forest 1	State of grass, nonflammable
2	green	Forest 2	State of shrubs, nonflammable
3	dark green	Forest 3	State of complete forest, flammable
4	yellow	Fire 1	The hottest state, fire spreads to the surrounding forest 3
5	orange	Fire 2	State of lower temperature, fire does not spread
6	red	Fire 3	State of fire declines, fire does not spread

Example results are shown in Fig. 7.11. Here, `cool1`= 1.0, `cool2`= 0.8, `cool3`= 0.8, `grow1`= 0.2, `grow2`= 0.1, `grow3`= 0.1, `diffus`= 0.8, `fire`= 2.0^{-5} parameters have been adopted.

Figure 7.12 shows a Swarm simulation of a forest fire, where

- (a) A fire started from the left end. Red, green, and black squares indicate fire, trees, and empty space, respectively. Large black areas represent areas where trees were burned.

- (b) Result when $F = 0.55$. Fire extended a little to the right and then extinguished.

- (c) Result when $F = 0.58$. Fire extended to the right end, but the forest as a whole was not burned.

- (d) Result when $F = 0.59$. This is a marginal case.

TABLE 7.7: Parameters of forest fire.

Parameter	Meaning	Explanation
`grow1`	Growth 1	Probability of changing from state 0 (ash) to state 1 (Forest 1)
`grow2`	Growth 2	Probability of changing from state 1 (Forest 1) to state 2 (Forest 2)
`grow3`	Growth 3	Probability of changing from state 2 (Forest 2) to state 3 (Forest 3)
`cool1`	Fire extinction 1	Probability of changing from state 4 (Fire 1) to state 5 (Fire 2)
`cool2`	Fire extinction 2	Probability of changing from state 5 (Fire 2) to state 6 (Fire 3)
`cool3`	Fire extinction 3	Probability of changing from state 6 (Fire 3) to state 0 (ash)
`fire`	Ignition	Probability of igniting at state 3 (Forest 3) and changing to state 4 (Fire 1)
`diffuse`	Fire spread	Probability of state 4 (Fire 1) spreading to the surrounding four neighborhoods of state 3 (Forest 3)

- (e) Result when $F = 0.63$. Fire extended to the right end, and the forest as a whole was burned.

7.5 Segregation model simulation with Swarm

Another example of a critical value arises in the well-known segregation model by Thomas Schelling [105]. Schelling designed a virtual space consisting of agents of two colors. Each agent prefers neighbors of the same color as itself in the virtual space. Schelling investigated the macroscopic phenomena that arise from microscopic preference. Here, preference is the threshold percentage of how many neighbors should be of the same color. The following are the basic rules of Schelling's model.

- All agents belong to one of two groups, and have a preference value on neighbors of the same color.

- Agents calculate the ratio of neighbors of the same color.

- The agent stays if the ratio is above the preference value. Otherwise, the agent randomly moves to an empty space where the preference criterion is satisfied.

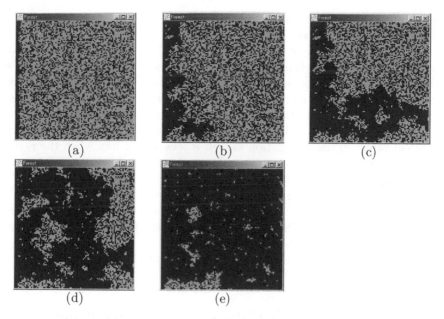

(a)   (b)   (c)

(d)   (e)

FIGURE 7.12: Examples of forest fires.

Schelling's results can be summarized as follows:

(1) Two colors segregated when the preference value exceeded 0.33.

(2) When the preference of one color was set higher than that of the other color, the color with the low preference spread out while the color with the high preference came together.

(3) The two colors also segregated when the preference criterion was changed to "the agent stays when there are three or more neighbors of the same color."

In summary, two colors ultimately segregated, and all agents wanted all neighbors to be of the same color. Schelling also found that the addition of relatively small changes to the preferences resulted in drastic changes to the macroscopic segregation patterns. In particular, the "prejudice," or bias in preference, of each agent and the segregation pattern are correlated. "Color blind" (pacifists with a preference criterion of zero) agents act to mix the colors. However, many critical values exist for these effects. The most significant finding is that society as a whole moves toward segregation even when agents only slightly dislike the other color (low preference threshold). Schelling's results highlighted many problems regarding racial discrimination and prejudice, and extensive research using improved versions of this model is still being conducted.

Schelling's results have raised many problems on "segregation" and "prejudice." A related issue is affirmative action (policies that actively counter discrimination to benefit under-represented groups based on historic and social environmental reasons).

An example is Starrett City in Brooklyn, NY, the largest federally assisted public rental housing scheme in the United States for middle-income residents that started construction in the mid-1970s. There was a requirement for residents that "limited the number of African-American and Hispanic residents to less than 40% of the total number of residents" to create a community where different races co-exist. The reason behind this policy is the theory of the "critical value" in the segregation model. Races do not mix when the ratio of whites becomes less than a certain value because of "white flight." Therefore, an attempt was made to build a stable community where people of various races live by keeping an appropriate balance of race and ethnicity. This policy was a success in one aspect: many families wanted to live in the development, and there was a waiting list (three to four months for white households and two years for black households).

However, whether these policies (e.g., preferential admission of minorities) are "righteous" is fiercely debated [103].

7.5.1   Swarm-based simulation of segregation model

This is an extension of the "Schelling" model with a "Church" in it. In this simulation, a maximum of three types of races can be introduced.

Rules for each individual's movement are as follows:

(1) Capacity per cell is one individual, and multiple individuals cannot exist.

(2) At each step, the candidate for the destination is selected randomly from a total of nine candidates (current position and the surrounding eight positions).

(3) If the satisfaction value obtained from the information of the eight surrounding neighbors is greater than the satisfaction value of the current position, then the move takes place.

(4) Basically, living in the church is not possible, but staying there is allowed if moving to some surrounding area is not possible.

The satisfaction value is calculated as follows:

- Satisfaction value increases by one if the agents of the same race exist in the surrounding eight places.

- Satisfaction value increases by one if a church exists in the surrounding eight places.

- Satisfaction value increases by three if the destination is a church.

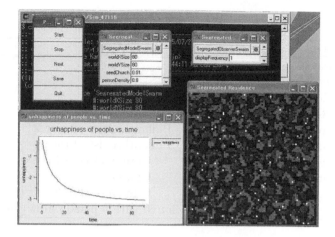

FIGURE 7.13 (See Color Insert): Schelling's simulation of the segregation model.

TABLE 7.8: Parameters of the separation model.

Parameter	Meaning
worldXSize	Extent of space (horizontal axis)
worldYSize	Extent of space (vertical axis)
seedChurch	Proportion of churches as compared to the extent of space
raceNum	Number of races (maximum value 3)
seedPhilan	Proportion of philanthropists (ratio against the total population)
personDensity	Population density against extent of space

- Satisfaction value increases by one if an agent exists in the surrounding eight places, in the case of a philanthropist.

The parameters inside the default file "seg.scm" are given in Table 7.8. An example of separation results is shown in Fig. 7.13. Parameters are: populationDensity= 0.8, seedChurch= 0.01, seedPhilan= 0.01. The meaning of cell colors is explained in Table 7.9.

7.6   Lattice gas automaton

Research conducted to model problems in fluids using cellular automaton (CA) has attracted interest since the 1980s. The lattice gas automaton (LGA)

TABLE 7.9: Meanings of the colors of the cells.

Color	Meaning
Brown	Race 1
Dark green	Race 2
Navy (deep blue)	Race 3
Red	Philanthropist of race 1
Green	Philanthropist of race 2
Blue	Philanthropist of race 3
Magenta	Church of race 1
Light blue	Church of race 2
Cyan	Church of race 3
White	The state of church with a person in it (for all races)

model is used to simulate fluids, where particles are placed on a lattice and then move to a neighboring lattice or collide with other particles based on predetermined rules at specified time units. The direction in which the particle moves after a collision is defined by simple rules.

The first LGA model was proposed in 1973 by Hardy, Pomeau, and de Pazzis, and thus is called the HPP model [41A]. Collision and scattering are repeated on a square lattice in this model. Only head-on collisions are considered, and particles scatter and change direction by 90° as in the following picture.

$$
\bullet \; \rightarrow \; \leftarrow \; \bullet \; \Longrightarrow \quad
\begin{array}{c} \uparrow \\ \bullet \\ \\ \bullet \\ \downarrow \end{array}
\tag{7.10}
$$

$$
\begin{array}{c} \bullet \\ \downarrow \\ \\ \uparrow \\ \bullet \end{array}
\quad \Longrightarrow \; \leftarrow \; \bullet \; \bullet \; \rightarrow
\tag{7.11}
$$

Note that the number of particles and the momentum are conserved in this model. Fig. 7.14 is an example of a collision process in the HPP model. However, false physical quantities and anisotropy of stress tensors (pressure) become issues in this model. This arises from a biased reaction because no reactions happen in the diagonal direction on the square lattice. Simulations on such models are in conflict with physical phenomena such as that described by the Navier–Stokes equations.

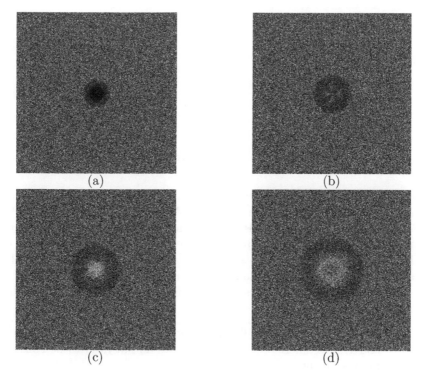

(a) (b)

(c) (d)

FIGURE 7.14: An example of a collision process in the HPP model ((a)→(b)→(c)→(d)).

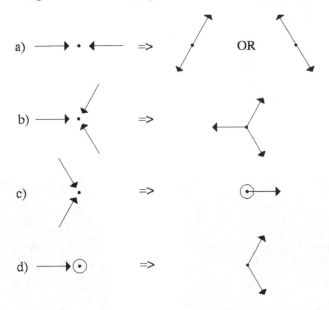

FIGURE 7.15: Collision process in the FHP model.

In 1987, Frisch, Hasslacher and Pomeau proposed an LGA method on a triangular lattice to resolve these issues [37A]. This is known as the FHP model. A collision process in the FHP model is shown in Fig. 7.15. The FHP method is used to simulate the flow of fluids around obstacles, for instance. This method is also used to model phase separation, phase transition, microemulsion by surfactants, heat flow, reaction, and diffusion.

The following advantages are obtained by analyzing fluids in the LGA method.

(1) Simulation is numerically stable because particles can take states of 0 and 1 only. Therefore, rounding errors do not need to be considered as in experiments.

(2) Reaction rules are very simple because the model is a repetition of collisions and diffusion.

(3) Each particle reacts independently; therefore, simulation can be highly parallelized.

(4) Complex boundary conditions can be set, making this suitable for simulation of complex systems.

However, the following disadvantages are known in the LGA method.

(1) Visualization is difficult because particles can take one of 0 and 1 states only. In other words, statistical procedures are necessary to obtain more relevant physical quantities.

FIGURE 7.16: Simulation examples of rain drops using the LGA method.

(2) Non-physical elements may appear.

(3) Dynamic parameters such as temperature are difficult to handle.

The lattice Boltzmann method incorporating probabilistic generation rules has recently been used in simulations of physical systems to overcome these drawbacks. Figure 7.16 is an example of simulation using the LGA method. Numerous videos of LGA simulations are also available online (see [30], for example).

7.6.1 LGA simulation with Swarm

LGA simulation by Swarm is shown in Fig. 7.17. Here, gas particles collision by the HPP model is implemented. The particles flow into the hollow (cavity) part from the horizontal and vertical directions, and their diffusion can be observed. As mentioned earlier, note that, since the collisions in the diagonal direction in a square lattice (grid) are not taken into account, there is a deviation in the reaction. Due to this limitation, false physical quantities and anisotropy of the stress tensor are caused.

Next, we show an example of simulation of fluid according to the lattice Boltzman law, created in Swarm. The lattice Boltzman law is an extension of the LGA law and has also been used extensively in fluid analysis. The LGA law describes by 0, 1 (integer number) the presence or absence of particles moving in some direction. However, in the lattice Boltzman law, the particle is represented by the local mean distribution function (real number), and deals with an equation of motion for the distribution function. In the lattice Boltzman law, by adjusting the coefficients of the distribution function, it can be implemented to eliminate non-physical elements. Therefore, a square lattice can also be used. Figure 7.18 is an example of the simulation. The gray color part represents a slab. For flow velocity, 0 is black, and becomes bigger

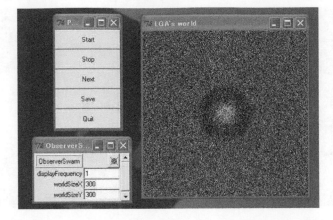

FIGURE 7.17: LGA simulation with the HPP model.

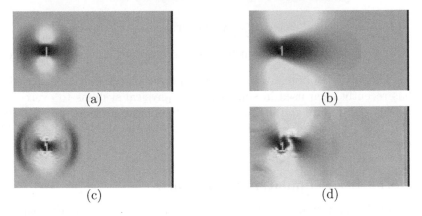

FIGURE 7.18 (See Color Insert): An example of simulation using the LGA method.

as it moves toward white from red. The number of lattice points is 50×100. By initial conditions, for all lattice points, a uniform velocity $\vec{u} = (0.1, 0)$ and a density of $\rho = 1.0$ have been specified. In the inflow boundary, uniform flow is assumed, and the distribution function is obtained by extrapolating the velocity of the outflow boundary.

7.7   Turing model and morphogenesis simulation

"Morpho" means "shape," and "genesis" means "generation." Alan Turing believed that the morphogenesis of organisms could be explained by the re-

(a) various shells	(b) chambered nautilus
	(@PNG in 2005)

FIGURE 7.19 (See Color Insert): CA patterns found on shells.

action and diffusion of morphogens, which are hypothetical chemicals, and he proposed the model described below. It is very interesting that von Neumann and Turing, pioneers in computer science, were exploring models of life phenomena in the early days of computers.

Two morphogens, X and Y, activate and inhibit, respectively. X and Y are governed by the following reaction and diffusion equations. x, y are the concentrations of X and Y; f, g are the generation rates of X and Y; and D is the diffusion coefficient.

$$\frac{\partial x}{\partial t} = f(x, y) + D_x \nabla^2 x, \tag{7.12}$$

$$\frac{\partial y}{\partial t} = g(x, y) + D_y \nabla^2 y. \tag{7.13}$$

The first term on the right-hand side is a generation term from a chemical reaction, and the second term represents movement by diffusion. These are called reaction–diffusion equations.

Using this model, Turing hypothesized that stable patterns would form if the evolution of the inhibitor is slower than that of the activator ($\frac{\partial x}{\partial t} > \frac{\partial y}{\partial t}$), and the diffusion of the inhibitor is faster than the activator ($D_y y > D_x x$). In fact, the formation of various patterns can be simulated by changing $f(x, y)$ and $g(x, y)$. The following are simulations of the Turing model using cellular automata (CA).

Kusch and Markus defined the CA rules for reaction and diffusion as shown below, and simulated the Turing model [77]. These are one-dimensional CA to reproduce patterns such as those found on shells and animal fur. Figure 7.19 shows a well-known pattern on a shell. The objective here is to generate such patterns.

Each cell has two variables, $u(t)$ and $v(t)$, corresponding to the amount of activator and inhibitor, respectively. $u(t)$ may be 0 or 1, where 0 is the dormant state (white) and 1 is the activated state (black). $u(t)$ and $v(t)$ transition to $u(t+1)$ and $v(t+1)$ through two intermediate steps, each according to the following rules.

(1) If `v(t)>=1`, then `v1=[v(t)(1-d)-e]`,
    else `v1=0`
(2) If `u(t)=0`, then `u1 = 1` with possibility p,
                and `u1 = 0` with possibility 1-p.
    else `u1=1`
(3) If `u1=1`, then `v2=v1+w1`,
    else `v2=v1`
(4) If `u1=0` and `nu>{m0+m1*v2}`, then `u2=1`,
    else `u2=u1`
(5) `v(t+1)={<v2>}`
(6) If `v(t+1)>=w2`, then `u(t+1)=0`,
    else `u(t+1)=u2`

Here, {} indicates the closest integer, < > is the average within distance `rv`, and `nu` is the number of activated cells within distance `ru`. (1) expresses the decrease in inhibitors per time step. In particular, a linear decrease is observed with $d = 1$ and $e = 1$, and an exponential decrease with $0 < d < 1$ and $e = 0$. Dormant cells are activated at a fixed probability according to (2). (3) shows that activated cells emit inhibitors. (4) states that a cell is activated if the number of activated cells within a distance (`nu`) is larger than the linear function (`m0+m1×v2`) of inhibitors (`v2`). This is the description of diffusion of activators. (5) means that the inhibitor becomes the average within the distance `rv`, showing how inhibitors diffuse. (6) states that a cell becomes inactive if the inhibitor mass is larger than the constant value.

Figure 7.20 shows the results of experiments along with the parameters of $d = 0.0, e = 1.0$, and initial probability `InitProb=` 0.0. Other parameters are shown in Table 7.10. Note that branching and interruption as seen in the figures are difficult to reproduce in differential equation based models, but are easily reproduced in CA models by using appropriate parameters. Results (e) and (h)–(k) correspond to Class IV described in Section 7.1.

In addition to organisms, waves from chemical reactions form patterns that have a rhythm. The Belousov–Zhabotinsky reaction is one representative example. The biologist Belousov from the former USSR discovered this reaction in the tricarboxylic acid cycle, an important reaction cycle in the energy metabolism of organisms.

7.7.1   Simulation of morphogenesis by the Turing model

To see the time evolution of one-dimensional cellular automata (CA), PatternSpace (`discrete2d`) is drawn. In the case of implementation of two-dimensional CA like the life game, it is good to use `DblBuffer2d` having two buffers. The CA used here has 10 parameters. By registering them in the

FIGURE 7.20: Turing model simulation results (see Table 7.10 for the parameters).

TABLE 7.10: Parameters for Turing model simulation.

(a)	ru=1,rv=17,w1=1,w2=1,m0=m1=0,p=0.002
(b)	ru=16,rv=0,w1=8,w2=21,m0=0,m1=1,p=0.002
(c)	ru=2,rv=0,w1=10,w2=48,m0=m1=0,p=0.002
(d)	ru=1,rv=16,w1=8,w2=6,m0=m1=0,p=0.002
(e)	ru=1,rv=17,w1=1,w2=1,m0=m1=0,p=0.002
(f)	ru=3,rv=8,w1=2,w2=11,m0=0,m1=0.3,p=0.001
(g)	ru=1,rv=23,w1=4,w2=61,m0=m1=0,p=0,d=0.05,e=0,initProb=0.1
(h)	ru=3,rv=8,w1=2,w2=11,m0=0,m1=0.3,p=0.001
(i)	ru=3,rv=0,w1=5,w2=12,m0=0,m1=0.22,p=0.004,d=0.19,e=0.0
(j)	ru=2,rv=0,w1=6,w2=35,m0=0,m1=0.05,p=0.002,d=0.1,e=0.0
(k)	ru=1,rv=2,w1=5,w2=10,m0=0,m1=0.3,p=0.002

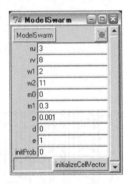

FIGURE 7.21: Parameters probe for the Turing model.

probe, they are made to be changed interactively (Fig. 7.21). As described earlier, parameters can be changed during execution by putting the values in the window (do not forget to input "return"). Note that in this simulation, the "`initialize`" method should be called for this setting.

7.8   Simulating percolation with Swarm

Let us consider a model of an infinite number of trees placed on lattice points where one tree is infected with a disease. This disease infects a neighboring tree with a certain probability p. The disease should stop spreading at some point if p is small, whereas it spreads infinitely if p exceeds a certain probability. Therefore, this model undergoes a phase transition at a certain value of the variable p. The value of p at the phase transition point is called the "critical probability" p_c. This model can be simplified as described in the next paragraph.

Suppose there is an infinite number of lattice points, and the probability that each lattice point is painted black is p. The black points are sparsely distributed on a plane when the probability p is low. However, there will be many chunks or clusters of lattice points painted black when the probability p becomes sufficiently high. An infinitely extending cluster will appear when the probability p exceeds the critical probability p_c. Such an infinitely extending cluster is called a percolated cluster, and the state where a percolated cluster exists is called a percolated state.

Percolation is a model for discussing the phase transition leading to percolated clusters.

We put points in a two-dimensional square lattice randomly. Adjacent points are considered connected. As the points increase, large clusters of connected points are formed. This is one of the models of the phenomenon called percolation.

When the space is infinitely large and the points exceed a certain density, it is well known that the size of the cluster also increases indefinitely. The probability of arbitrarily selected grid (lattice) points becoming a part of the infinitely large cluster is called the percolation probability. This is the ratio of the number of points in the largest cluster to the total number of lattice points.

Here, we will examine the percolation probability in the case of finite space. Starting with nothing, the gird is filled randomly, and the determination of the size of the cluster is repeated. At that time, since the process of calculation of clustering is computationally expensive, it should not be repeated each time a grid/lattice is filled. Preparing two schedules and separately registering grid filling action and clustering action can easily accomplish this. The whole structure looks like Fig. 7.22.

If we look at "`patternSpace`," it can be observed that there is a point when a small cluster (dense portion) suddenly increases to become the largest (Fig. 7.23).

The percolation probability changes as in Fig. 7.24. It is very small (0 in the case of infinitely large) until a critical value (0.5927 in the case of a two-dimensional square grid) called the threshold percolation, and increases suddenly beyond that. Here, the transition of the number of clusters is also graphed (divided by 15% of the number of grids). It can be seen that it rapidly decreases with the increase in percolation probability.

Histograms are also provided in Swarm. By specifying the targeted objects collection and the methods to retrieve, the values can easily create them (Fig. 7.25). `EZBin` displays the histograms. `EZBin` is the wrapper about the objects, and specifies the number of the "`bin`" and the interval between the displays. The following steps set this object:

(1) Generation of instance

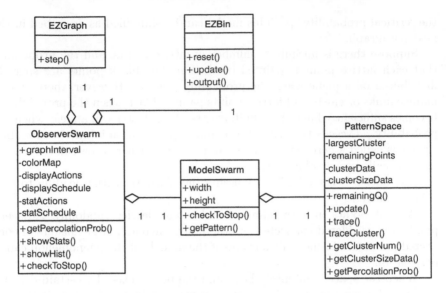

FIGURE 7.22: The whole structure for percolation.

FIGURE 7.23: patternSpace: the dark portion is the largest cluster.

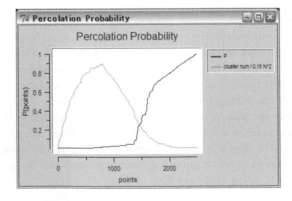

FIGURE 7.24: Percolation probability.

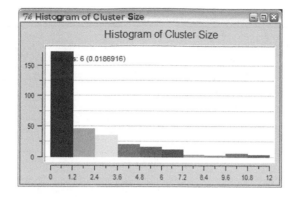

FIGURE 7.25: Histograms in Swarm.

(2) Setting of title

(3) Setting of collection

(4) Setting of "`selector`" of probe

(5) Setting of number of "`bin`"

(6) Setting of upper and lower limits of display

7.9   Silicon traffic and its control

The following is an explanation of traffic modeling. Nagel and Shreckenberg modeled traffic congestion using a one-dimensional CA [89]. In their model, each cell corresponds to a position where cars can pass, and each cell can take one of two states (one or no car is in the cell) in every time step. Every car moves at a characteristic speed (integer value from 0 to v_{max}). The speed of the cars and the state of the cells are updated based on the following rules.

Acceleration Increase the speed by one unit ($v := v+1$) if the speed of a car v is less than v_{max} and the distance from the first car ahead is greater than $v + 1$.

Deceleration Decrease the speed of a car at i to $j - 1$ ($v = j - 1$) if the first car ahead is located at $i + j$ and $j \leq v$.

Random number Decrease the speed of all cars by 1 (if larger than 0) at possibility p ($v := v - 1$).

Move Move all cars at their respective speed v.

space (road)

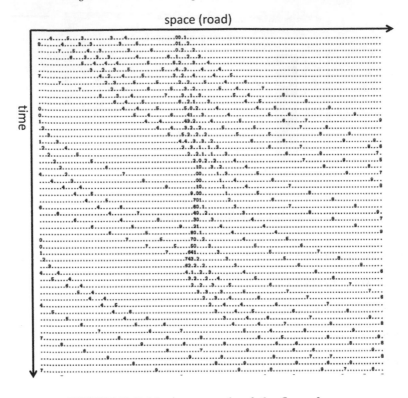

FIGURE 7.26: An example of the flow of cars.

The most important parameter in this simulation is the density of cars ρ, defined as

$$\rho = \frac{\text{total number of cars}}{\text{total number of cells}} \tag{7.14}$$

Figure 7.26 is an example of the flow of cars with $v_{\max} = 5$ and $\rho = 0.1$. Cars move from left to right, and the right end is connected to the left end. The numbers indicate the speed of the cars, and the dots show cells with no cars. Consecutive cars at speed 0 at the center indicate traffic congestion.

Nagel and Shreckenberg conducted various experiments and found that the nature of congestion changes at $\rho = 0.08$. Note that congestion is measured by the average speed of all cars. Thus, the critical value for this traffic simulation is $\rho = 0.08$.

7.9.1   Simulating traffic jams with Swarm

Nagels and Shreckenberg performed the modeling of traffic jams using a one-dimensional lattice. In their model, each cell corresponds to the vehicle's passage, at each time-step, and the cells take one of the two states (vacant or

FIGURE 7.27 (See Color Insert): Simulating a traffic jam (with Slow-toStart).

occupied by one vehicle). All the vehicles move with a specific speed (integer number, 0 to v_{max}). This model was visualized and implemented on Swarm. To see the time change of one-dimensional cellular automaton (CA), it is drawn on `PatternSpace` (`Discrete2d`). To implement a two-dimensional CA like the life game, it is good to use `DblBuffer2d`, which has two buffers. For example, the movement of a car with $v_{max} = 10$, $\rho = 0.1$ is shown in Fig. 7.27. The vehicle moves in the right-hand direction, and the right end and the left end are connected. The top row shows the current road conditions, and the time passed as we move down (displayed for 200 steps). In Swarm, the vehicles' speed is represented by shades of colors. In fact, green becomes darker with the increase in speed, and the red becomes darker as the speed decreases (as set in `colorMap`). A chunk of red color is the point of occurrence of a traffic jam, and the black part shows a cell without a vehicle. In Fig. 7.27, the occurrence of a series of vehicles with speed 0 in the middle is a traffic jam. We can see that the traffic jam moves forward with time.

In this model, SlowtoStart has been introduced to realize the effect of inertia. This is a rule that says once the vehicle has stopped, it starts moving after 1 time step even if the front is open to move. This is considered important to bring the model closer to actual metastability. If SlowtoStart is used, it takes time to accelerate, which leads to more stationary vehicles and worsens the traffic jam (Fig. 7.27). On the other hand, without SlowtoStart, the line of the stationary vehicles will not elongate unless there is a slowdown due to a random number, and therefore as a result, the traffic jam is eased (Fig. 7.28).

Recently ASEP (Asymmetric Simple Exclusion Process) has been studied extensively as a model of traffic flow [99]. In this model, the maximum speed of each vehicle existing in each cell is taken as 1; if the cell in front is vacant, the vehicle moves to it with a probability p (stops with a probability $1 - p$). Let us take the inflow and outflow probabilities of new cars as α and β, respectively (in other words, right and left ends are not connected, and the number of vehicles is not fixed). Figure 7.29 shows the ASEP model simulation in Swarm. Parameters are $\alpha = 0.3$, $\beta = 0.9$, $p = 0.5$ for (a) free phase, $\alpha = 0.3$,

FIGURE 7.28 (See Color Insert): Simulating a traffic jam (without Slow-toStart).

(a) free phase　　　　(b) shockwave phase　　　(c) maximum flow phase

FIGURE 7.29: ASEP model simulation.

$\beta = 0.3$, $p = 0.5$ for (b) shockwave phase, and $\alpha = 0.6$, $\beta = 0.6$, $p = 0.5$ for (c) maximum flow phase. In this case, `colorMap` is set such that black points show cells with vehicles, and the white part shows vacant cells. As explained earlier, the topmost row is the current road situation. As we go down, it shows the time already passed. In the ASEP model, various mathematical analyses are done for the equilibrium state [99].

7.10　The world of Sugarscape

In this section, we discuss Sugarscape models and their simulation. Sugarscape was presented by Joshua M. Epstein and Robert Axtell as an experimental model environment for creating artificial societies. Sugarscape is based on the simulation of an unlimited number of ants walking around in search of sugar. The model can construct artificial societies with the aim of examining the mechanisms driving various social phenomena, and allows the

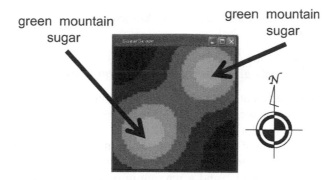

FIGURE 7.30: Two sugar (green) mountains in the Sugarscape model.

incorporation of concepts such as breeding, pollution, culture, combat, trade, and disease.

The notations and definitions regarding Sugarscape presented in this section are based on [33].

7.10.1 A simple Sugarscape model

Agents live in a space named Sugarscape, which is a two-dimensional square lattice toroidally connected at the top, bottom, and left and right edges. Each lattice point contains sugar, and the maximum quantity of sugar is set on a per-point basis. The amount of sugar increases at a predefined rate (cf. regeneration rules G_α and G_∞, which are described below). The agents collect sugar from lattice points, and the sugar at each point is restored to its maximum if not collected by agents for a short period of time. Sugar is thus maximized at all points at the initial iteration step.

Figure 7.30 shows an example Sugarscape model. Green mountains are located at the northeast (top right) and southwest (bottom left) corners of the lattice, where areas of the mountains with greater elevation represent larger amounts of sugar, with the maximum quantity at the peaks set to 4. In contrast, the areas toward the northwest and southeast corners eventually become completely devoid of sugar. Note that since the lattice is toroidally connected, when an agent crosses the right (left) edge, it reappears on the left (right) edge. We will now explain this simple environment for the initial model.

Initially, the quantity of sugar everywhere increases according to the simple rule described next [33, p. 23]. Various other rules for sugar growth have been devised, and the rule given here is extended in later sections.

Sugarscape growth rule G_α:

The sugar grows at a rate of α units per step, where the maximum growth is such that the upper limit of the quantity for that position is attained.

In particular, the following rule is known as the instantaneous growth rule [33, p. 27].

Sugarscape growth rule G_∞:
Sugar instantaneously grows to the maximum quantity for that position.

Two important agent features are their vision and metabolic rate. These features are different for each agent and can be genetically transferred from parent to child. Therefore, although the agents consume (burn) sugar at each step, the consumption of sugar depends on the metabolic rate. In the following experiment the metabolic rate for the population forms a uniform distribution with a minimum value of 1 and a maximum of 4. The vision of the agents is limited to the four directions of the lattice (up, down, left, and right), and the agents are unable to see across a diagonal. Furthermore, the distribution of vision is uniform, with a minimum of 1 and a maximum of 6, and an agent with a vision of 3 can see up to three units away in each direction.

The agents thus accumulate sugar while moving around in the Sugarscape, and each agent is capable of stockpiling an unlimited amount of sugar.

Agents move in accordance with the rule given below [33, p. 25], in which they process local information (such as the current amount of sugar at each position) within their field of view, and subsequently compute their order of preference for relocation.

Agent movement rule M:

- Survey the lattice in all four directions (up, down, left, and right) and search for positions within the field of vision containing the largest quantity of sugar and no other agents.

- If more than one such position exists, choose the nearest.

- Move to the new position.

- Collect all the sugar at the new position.

Agents can move only once at each step of the iteration, and the order in which agents move is random. When they arrive at the new position, their sugar reserve is increased by the amount of sugar at the new position minus the amount of metabolized sugar. Metabolized sugar cannot be accumulated, and if an agent's reserve becomes zero or negative, the agent dies of starvation and disappears from Sugarscape.

In the following discussion, if Sugarscape follows growth rule E and the agents follow movement rule A, the set of experimental conditions is denoted (E,A). For example, if sugar is replenished by the instantaneous regeneration rule and the agents move in accordance with rule M, then the experimental conditions are given by (G_∞,M).

Let us conduct a simulation with the (G_1, M) rule set. Applying these rules to a group of agents distributed at random, agents located at positions with

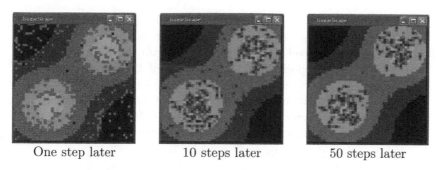

| One step later | 10 steps later | 50 steps later |

FIGURE 7.31: Agent aggregation in (G_1, M).

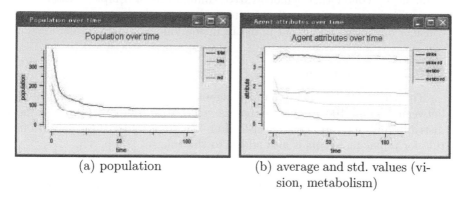

(a) population (b) average and std. values (vision, metabolism)

FIGURE 7.32: Population and feature changes in (G_1, M).

little sugar exhaust their food and die, and only agents in positions relatively rich in sugar survive. Moreover, since agents rely on their field of view to discover and move to positions rich in sugar, the agents aggregate at the two mountains of sugar after 50 steps (Fig. 7.31).

After 50 steps, the population of agents has been reduced to about 100 from a value of 400 at the start of the simulation, and although the average metabolic rate for all agents is initially 2.5, this value converges to about 1. However, no improvement in vision is found even after 100 steps. Thus, vision can be considered as not being of prime importance in this environment, or perhaps vision is already sufficiently advanced in this case (Fig. 7.32).

Next, let us experiment with the (G_∞, M) rule set. In this environment, the survival rate is low for agents with high metabolic rates and poor vision. Therefore, such agents die while the remaining agents settle at the most suitable locations and form a stable state. As a result, it can be observed that rather than aggregating at the peaks of the two mountains, the agents arrange themselves in a formation resembling contour lines along the mountain ridges (terraces) of Sugarscape [33]. Epstein and Axtell explained this phenomenon as originating from the vision of the agents:

Specifically, suppose you are an agent with vision 2 and you are born on the terrace of sugar height 2, just one unit south of the sugar terrace of level 3. With vision 2, you can see that the nearest maximum sugar position is on the ridge of the sugar terrace of height 3, so, obeying rule **M**, you go there and collect the sugar. Since there is instant growback, no point *on* the level 3 sugar terrace is an *improvement*; and with vision of only 2, you cannot see the higher terrace of sugar level 4. So you stick on the ridge [33, p. 28].

This description suggests that while Sugarscape constitutes a simple environment, by conducting demonstrative simulations of complex systems, Sugarscape is an exceedingly effective tool for studying human society. Let us consider one such experiment in detail in the following paragraphs.

7.10.2   Life and birth

To increase the model dynamics, agents are born with a certain lifespan between 60 and 100 iterations that is selected at random from a uniform distribution. Since all agents perish after 100 steps under these conditions, a rule for replacing agents is also added to the rule set [33, p. 33].

Agent replacement rule $R[a, b]$
An agent that dies is replaced by a new agent with an age of 0 and a random set of genes, lifespan (with an interval set as $[a, b]$), initial reserve, and location.

Explicitly, since a single dead agent is replaced by a single agent generated at random, the population is maintained at 400. The implementation of this rule results in small fluctuations in the average values of the agents' attributes owing to the addition of the random attribute values of the new agent. Under these conditions, the metabolic rate decreases rapidly, whereas vision remains almost unchanged (Fig. 7.33). Additionally, although the agents accumulate the surplus sugar and thus possess considerable reserves, the average stored quantity appears to settle around 30 due to the limited lifespan of the agents. The maximum size of a reserve is about 100 (Fig. 7.34).

7.10.3   Breeding

The last basic setting is a breeding rule. Agents are divided into males and females, and adjacent agents of opposite sex breed and produce offspring. However, for successful breeding, the following conditions must be satisfied.

- Age limit: both males and females can start breeding when they are between 12 and 15 steps of age, and breeding stops between 40 and 50 for females and between 50 and 60 for males. Each agent is assigned

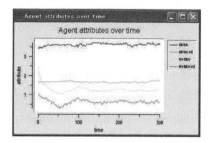

FIGURE 7.33: Features (vision, metabolism) in $(G_1, \{M, R_{[60,100]}\})$ (average and std. values).

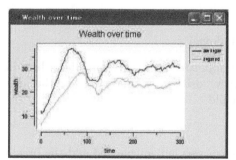

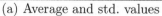

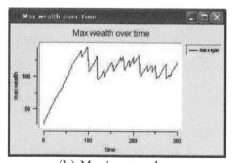

(a) Average and std. values   (b) Maximum values

FIGURE 7.34: Wealth in $(G_1, \{M, R_{[60,100]}\})$.

TABLE 7.11: Potential child genotypes [33].

Vision	Metabolic rate	
	m	M
v	(m, v)	(M, v)
V	(m, V)	(M, V)

a lower and upper breeding age taken from uniform distributions over these two intervals.

- Reserve: to become a parent, an agent must possess at least as much sugar in its reserve as at its birth.

In addition, the father and mother of a child each give the child half of the quantity of sugar that they possessed at birth. In other words, a child is born with a quantity of sugar equal to the sum of the amounts donated by its parents.

In the breeding process, each attribute (e.g., metabolic rate, vision, and maximum lifespan) is randomly inherited by the child from one of its parents. Therefore, rather than being generated at random, the attributes of a newly born agent represent a uniform crossover of the genetic elements of its parent agents. Taking metabolic rate and vision as an example, if the genotype of one parent is (m, v) and the genotype of the other is (M, V), the child can obtain one of four genotype combinations with equal probability (Table 7.11): (m, v), (M, v), (m, V), and (M, V). The following rule is adopted for breeding agents [33, p. 56].

Agent breeding rule S:

- Select an adjacent agent at random.

- If the selected agent is of the opposite sex and is capable of breeding, and if there is at least one empty position for child next to either agent, then the agents produce one child.

- Repeat for all agents.

In an experiment where the lifespan was left unchanged and the breeding rule was substituted for the replacement rule, the agent population initially decreased in barren regions and then increased rapidly through breeding in rich regions, before eventually settling at around 700. Here, average values for both the vision and metabolic rate of the agents changed; the metabolic rate decreased even more rapidly as compared with the case without crossovers, and the vision improved (Fig. 7.35). The reason for this improvement in sight, in addition to evolution, is that the agents experienced difficulty in finding free space as the population increased, and therefore the importance of having a

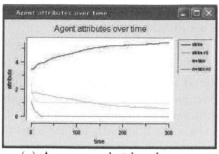
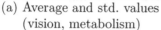

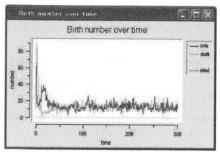

(a) Average and std. values (vision, metabolism)

(b) Numbers of births and deaths

FIGURE 7.35: Features in $(G_1, \{M, R_{[60,100]}, S\})$.

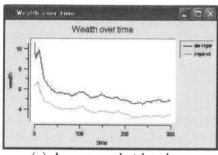

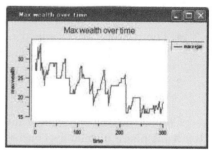

(a) Average and std. values

(b) Maximum values

FIGURE 7.36: Wealth in $(G_1, \{M, R_{[60,100]}, S\})$.

wide field of vision also increased. Furthermore, the average reserve decreased to 6 as a result of agents donating part of their reserve to their newly born child, and the maximum reserve decreased to about 15 (Fig. 7.36).

7.10.4 Environmental changes

In this section, we observe the behavior of the agents when their environment is changed. The basic rules in the following discussion are the breeding rule from Section 7.10.3 and the rule that says lifespans are uniformly distributed over the interval 60–100.

7.10.4.1 Nutritive ratio

Let us lower the nutritive ratio of sugar to 40%; in other words, only 4 out of 10 collected units of sugar can be digested, which is equivalent to reducing the consumption of sugar to 40%. When the experiment is conducted under these conditions, the sugar collected by the agents is unable to cover the energy needs of their metabolisms. As a result, the agents exhaust their reserves, and the entire population perishes at the 25th step. The minimum nutritive ratio of sugar at which the agents survive is about 50%.

7.10.4.2   Alternating seasons

Next, alternating seasons are introduced as a form of dynamic environmental change. Taking the upper half of Sugarscape as the northern hemisphere and the lower half as the southern hemisphere, the hemispheres alternate between summer and winter every 50 steps such that both hemispheres have opposing seasons. The sugar yield in winter is 1/8 of that in summer. The season rule is summarized as follows [33, p. 45].

Rule for season-dependent sugar growth $S_{\alpha\beta\gamma}$:

- The season is set to summer in the upper (northern) half and winter in the lower (southern) half of Sugarscape.

- The seasons are swapped after a period of time α (summer becomes winter and winter becomes summer).

- Sugar grows at a rate of γ units per step in summer, and a rate of γ units per β steps in winter.

Conducting the experiment under these conditions, the agents in the winter hemisphere ultimately consume all of the sugar, and their population is drastically reduced. Only agents located around the equator can survive by migrating to the summer hemisphere. However, once the seasons are swapped after 50 steps, the agents that have been consuming generous amounts of sugar during the summer suddenly face a severe winter (Fig. 7.37), and agents die in large numbers if they do not reach the summer hemisphere. Agents that successfully relocate produce offspring and increase in number (Fig. 7.38).

If the seasons are swapped an odd number of times, values of the average vision and metabolic rate both improve, whereas with an even number of swaps these attributes degenerate slightly. Agents with superior features adapt well to the change in season and gather together inside the summer hemisphere. However, since after an odd number of swaps the season in that hemisphere is winter, a large number of agents die, and the remaining agents in the opposite hemisphere produce offspring and increase the population. Therefore, the effect of location-based selection is stronger than that of superior or inferior features, resulting in an overall degeneration of the attributes. Furthermore, the average reserve size increases rapidly immediately after the swap, owing to the small number of agents refilling their reserves in the summer hemisphere (Fig. 7.39).

7.10.4.3   Generation of pollution

When the climate alternates, the environment exerts a unidirectional influence on the agents. In contrast, in this section we introduce pollution as an

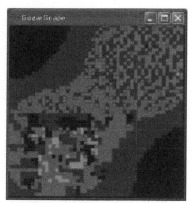 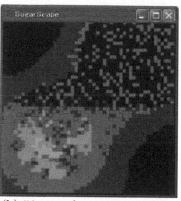

(a) 25 steps later; winter in the lower hemisphere (southern)

(b) 50 steps later; winter in the upper hemisphere (northern)

FIGURE 7.37: Agent aggregation due to seasonal variation: $(S_{50,8,1}, \{M, R_{[60,100]}, S\})$.

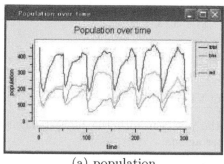 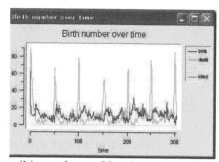

(a) population

(b) numbers of births and deaths

FIGURE 7.38: Population changes due to seasonal variation: $(S_{50,8,1}, \{M, R_{[60,100]}, S\})$.

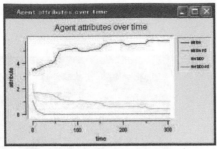
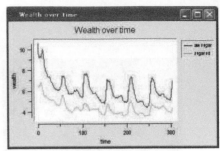

(a) average and std. values (vision, metabolism)

(b) average and std. values (wealth)

FIGURE 7.39: Feature and wealth changes due to seasonal variation: $(S_{50,8,1}, \{M, R_{[60,100]}, S\})$.

example to show agents affected by their environment. Upon collecting and digesting sugar, an agent generates a corresponding amount of pollutant at its position, which makes it difficult for the agent to continue living there. The rule for generation of pollution is as follows [33, p. 47]:

Pollution generation rule $P_{\alpha\beta}$:

- Agents generate $\alpha \cdot s$ units of pollutant upon collecting s units of sugar (production waste) and $\beta \cdot m$ units of pollutant upon digesting m units of sugar (consumption waste).

- The total amount of pollutant $p(t)$ at time t for a given position is the sum of the production waste, the consumption waste, and the existing amount of pollutant, which can be expressed as

$$p(t) = p(t - 1) + \alpha \cdot s + \beta \cdot m. \tag{7.15}$$

The experiment below is conducted with $\alpha = 1$ and $\beta = 1$, and the movement rule is adjusted as follows [33, p. 48]:

Movement rule M adjusted to account for pollution:

- As vision permits, search for the position with the highest sugar-to-pollutant ratio and with no other agents.

- If more than one such position exists, choose the nearest.

- Move to the selected position and collect all of the available sugar.

In addition, the pollutant is dispersed with a fixed ratio in accordance with the following rule [33, p. 48]:

Pollution dispersal rule D_ϕ:

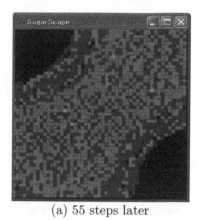

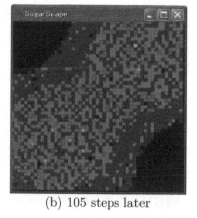

(a) 55 steps later   (b) 105 steps later

FIGURE 7.40: Agent aggregation due to pollution: $(\{G_1, D_1\}, \{M, P_{1,1}\})$.

- Calculate the input pollutant per interval ϕ for each cell, where the input is the averaged pollution level of the cells above, below, left, and right of the cell of interest (the von Neumann neighborhood).

- The input pollutant is set as the new pollution level at that cell.

Here, we take $\phi = 1$, and pollution generation and dispersal begin at the 50th and 100th step, respectively. This model thus matches historic processes where pollution increases in response to industrialization, after which efforts to preserve the environment expand as a result of technological development. When the experiment is conducted under these conditions, the moment that pollution generation begins, agents move away from the mountains and distribute themselves over a wide area. However, once pollution dispersal commences, agents return to the mountains (Fig. 7.40).

Although the population decreases temporarily when pollution is generated, curiously it recovers even before pollution dispersal initiates and the population reaches higher levels (about 800) than in the no pollution case. When pollution dispersal begins, the population returns to the typical level of about 700. This process occurs since, under normal conditions, the agents are preferentially distributed around the centers of the two mountains. However, pollution widens the agents' distribution and relaxes this bias; in other words, sugar collection becomes more efficient, covering a wide area that reaches the bases of the two sugar mountains, and the overall accommodated population increases.

When pollution is generated, agents' vision deteriorates on average, whereas their average reserves increase. The pollution-driven movement of the agents causes a decrease in population density. Thus, agents can collect sugar from their neighborhood without interference from other agents and this in turn decreases the importance of vision. This tendency is lost when pollution dispersal begins, and the agents resume their usual state (Fig. 7.41).

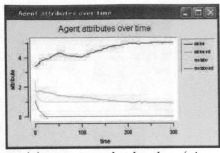
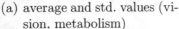
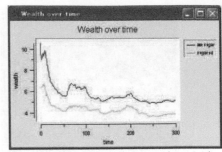

(a) average and std. values (vision, metabolism)   (b) average and std. values (wealth)

FIGURE 7.41: Feature and wealth changes due to pollution: $(S_{50,8,1}, \{M, R_{[60,100]}, S\})$.

7.10.5 Introduction of culture

In this section, we introduce interaction between agents. Each agent is provided with a fixed-length character array composed of 0s and 1s, which represents the culture of that agent. Agents belonging to a specific culture form a tribe, where the definition of cultures and tribes is as follows [33, p. 73]:

Cultures (tribes) defined by majority rule based on tags:

- Each agent carries a tag (a string array containing 0s and 1s).

- Agents whose tags contain more (less) 1s than 0s belong to the blue (red) culture.

Through breeding, parents produce offspring that inherit a random mixture of the parents' culture tags. When an attempt is made to divide the agents by their color, the two tribes mix at random. Furthermore, there is a gradation in nuance depending on the proportion of 1s in the tags, which indicates that although only two tribes (red and blue) are specified, each tribe includes a stepwise cultural division.

In general, the current features of agents do not change upon the inheritance of culture tags. Therefore, we introduce a new rule that enables tags to be propagated by flipping 0s and 1s in the tags of adjacent agents [33, p. 72].

Culture propagation rule (tag flip) K:

- A tag-flipping agent selects random positions in the tags of each of its neighboring agents.

- If the values at the selected positions are equal to those in the tag of the tag-flipping agent itself, they remain unchanged; otherwise they are flipped to match those of the tag-flipping agent.

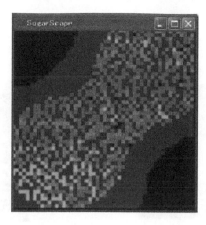

FIGURE 7.42: Agent aggregation with culture propagation (100 steps later): $(G_1, \{M, K\})$.

The tag length is taken as 3 in the following experiment. Although features, such as vision and metabolic rate, correspond to genes that can be transferred only through congenital inheritance, culture tags are memes, transferrable through secondary contact [25]. Upon application of the culture propagation rule, instead of diffusing at random, the red and blue tribes become more segregated than previously (Fig. 7.42).

The preceding example exhibits the formation of groups from randomly mixed blue and red tribes. In contrast, the next example shows the process of diffusion of the two tribes when they are completely separated initially. The experiment commences after placing the red tribe in the lower left-hand corner and the blue tribe in the top right-hand corner, and the interaction between the two tribes at the center of Sugarscape is then observed. If culture propagation is not implemented, only a small number of agents are born with mixed culture through inter-tribal breeding (Fig. 7.43). However, culture propagation causes rapid diffusion of the two cultures, and a large number of agents with intermediate culture are born, advancing deep into both tribes (Fig. 7.44).

While cases exists where, as the iterations advance, one culture vanishes and all agents become approximately evenly distributed within a single tribe, other cases again give rise to the vanished tribe through culture propagation (or inheritance). For example, even though agents with tags 001 and 100 belong to the red tribe, if they meet and influence each other, both agents acquire tags of 101, and they then belong to the blue tribe. In this way, preservation of cultural diversity is possible.

When the length of the tag is increased from 3 to 7, agents with intermediate culture again appear in a stepwise manner; however, neither tribe is completely converted even after 200 steps (Fig. 7.45).

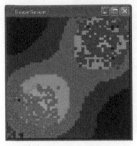 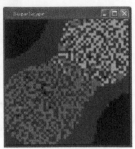

(a) Initial state	(b) 100 steps later

FIGURE 7.43: Diffusion process of two tribes: without culture propagation.

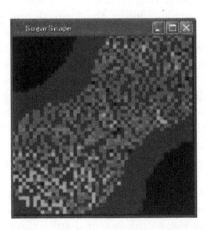

FIGURE 7.44: Diffusion process of two tribes: with culture propagation (100 steps later).

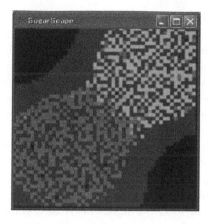

FIGURE 7.45: Diffusion process of two tribes: with culture propagation (no. of tags = 7; 200 steps later).

7.10.6 Introduction of combat

We now introduce combat between the two tribes, where each agent possesses a strength value dependent on its reserve. A strong agent can kill a weaker agent from the rival tribe and steal its reserve. The combat rule is summarized as follows [33, p. 83]:

Combat rule C_α:

- Survey the lattice in four directions (up, down, left, and right) as vision permits.

- Exclude cells occupied by members of the same tribe.

- If a cell is occupied by a weaker member of the other tribe (i.e., one with a smaller reserve), add a premium (of at most α) to the existing resource at the cell.

- Exclude cells occupied by stronger members of the enemy tribe (to avoid being attacked).

- Of the remaining cells, move to the nearest cell with the largest combined value of resource and premium.

- Collect the resource at the new location together with the reserve of the previous occupant and the premium α.

- The previous occupant of the cell dies.

When the combat rule is applied without culture propagation, an intriguing war unfolds between the two tribes (Fig. 7.46). The population graph shows that one of the tribes almost completely conquers the other, after which the low-population tribe, which was on the verge of extinction, suddenly recovers and reverses the combat. The population of the winning tribe is then reduced to almost complete annihilation. This phenomenon repeats indefinitely.

To explain this cycle, we consider Fig. 7.46 in greater detail. In this figure, in addition to division of the tribes by color, agents with large reserves (i.e., strong agents) are shown in a bright color, whereas agents with smaller reserves (weaker agents) are shown in a dark color.

At step 120, the blue tribe contains a large number of agents, and the mountain in the lower part of Sugarscape is inhabited only by blue agents. With no combatant, blue agents are unable to obtain a premium, and therefore their only means of support is to collect sugar from the land. Moreover, the relative amount of food per agent decreases, since their population density increases. Hence, the strength of the populous blue tribe is reduced. In contrast, only a small number of red agents survive at the mountain in the upper part of Sugarscape, and because they fight the blue agents, the local population density is lower, allowing for large quantities of sugar to be accumulated. Since the winner receives the sugar reserve of a defeated enemy, agents that are strong in combat become even stronger. Consequently, when a small number of red agents haphazardly win a continuous series of fights with the surrounding blue agents, they rapidly accumulate large amounts of sugar and become exceedingly powerful. These strong red agents then begin to systematically eradicate the weakened blue agents. As a result, the red agents successfully conquer the mountain in the lower half of Sugarscape at step 250. However, the red agents at the mountain in the upper half have already started to weaken, and the few blue agents struggling for survival suddenly accumulate large amounts of sugar, become powerful and expel the red agents. The cyclic combat history of the two tribes is formed in this way.

Next, we consider the addition of culture propagation, where agents are affected by the culture of the enemy tribe, and in certain cases even defect to the enemy side. Although dynamic combat can still be observed under these conditions (Fig. 7.47), fluctuations in population sizes are not as extreme as in the absence of cultural propagation. As expected, conversion can occur during combat and the less populous tribe recovers before an extreme bias in the distribution arises. Cases are rare where the less populous tribe is completely converted to the culture of the more populous tribe and the latter wins.

To observe the effects of combat, C_α is introduced at the 200th step of an experiment initially without combat. The blue and red agents naturally coexist in peace until the introduction of combat, at which point they swiftly separate into tribes and display dynamic behavior (Fig. 7.48). With the introduction of combat, the total population considerably decreases due to combat losses, and the average agent reserve increases following the theft of resources and the decrease in population.

While the same conditions were applied to both tribes in the preceding examples, finally we introduce differences in the personalities of the agents in

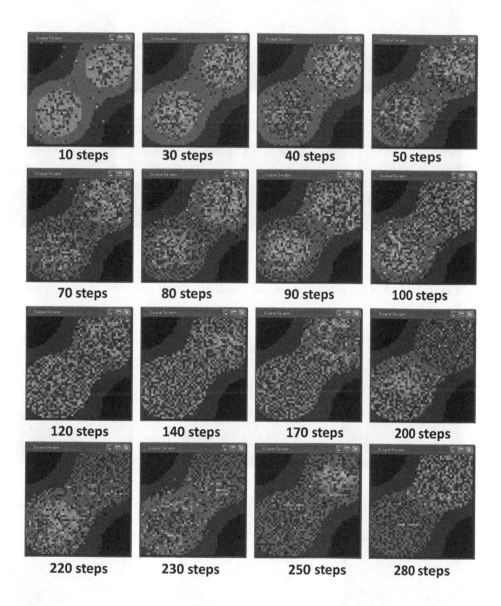

FIGURE 7.46: Introduction of combat (without culture propagation): (G_1, C_∞).

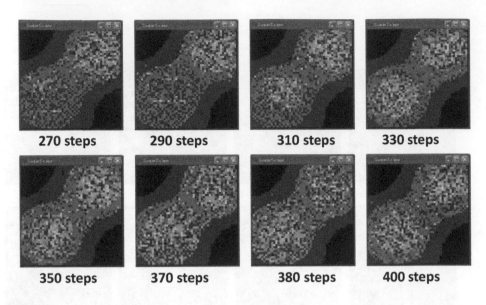

FIGURE 7.47: Introduction of combat (with culture propagation): (G_1, C_∞).

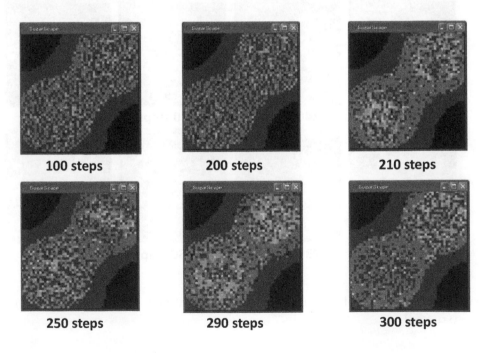

FIGURE 7.48 (See Color Insert): Combat introduced at the 200th step.

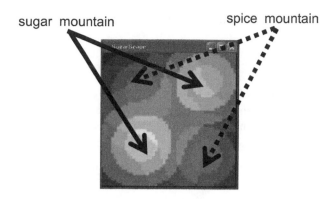

FIGURE 7.49: Introduction of spice.

both tribes. C_α contains a condition whereby an agent does not move toward a cell if a stronger agent from the rival tribe is in sight. Here, we define a new movement rule where an agent deliberately moves to a cell containing large amounts of food and a large premium, even if a strong enemy is in sight. The former behavior is regarded as being cautious, while the latter is regarded as being brave (or reckless).

When the experiment is conducted by taking the blue agents as brave and the red agents as cautious, the previously mentioned oscillation in the tribes' populations does not occur, and instead the population of the blue tribe increases and remains stable. The same process is observed when the roles are reversed and we have brave red agents and cautious blue agents. This result arises from the limited movement of the cautious agents due to fear of the enemy, in which they miss opportunities to consume large quantities of sugar or to steal resources. In this model, a brave personality proves to be more rewarding than a cautious one.

7.10.7   Introduction of trade

In this section, we introduce a second type of food (spice) in addition to sugar such that agents trade sugar and spice with neighboring agents whenever necessary. Thus, we attempt to create a bottom-up emergent market.

Agents consume both sugar and spice, and die if either is lacking. Spice also forms two mountains and grows at a predefined rate. The sugar mountains are shown in green and the spice mountains are shown in red (Fig. 7.49). Spice occupies the northwest and southeast corners of Sugarscape, which are poor in sugar. In the successive paragraphs, the metabolic rates of sugar and spice for each agent are uniformly distributed between 1 and 3, and metabolic rates are transferred independently to offspring.

We also introduce a welfare function, representing the cumulative reserve of sugar and spice. This welfare function is expressed in the form of a Cobb–Douglas function:

$$W(w_1, w_2) = w_1^{\frac{m_1}{m_T}} w_2^{\frac{m_2}{m_T}}. \qquad (7.16)$$

Here, w_1 is the amount of sugar, w_2 is the amount of spice, m_1 is the metabolic rate for sugar, m_2 is the metabolic rate for spice, and m_T is the sum of m_1 and m_2. This welfare function changes the movement rule for an agent at a given time [33, p. 100].

Movement rule, M, for an agent at a given time in the presence of multiple types of food:

- Survey the lattice in four directions as vision permits.

- Search for cells where the welfare function is maximized and no agents are present.

- If more than one such cell exists, choose the nearest.

- Move to the chosen cell and consume all resources.

With this rule, agents move toward cells that provide the optimal balance in satisfying their energy needs with respect to their metabolic rates of sugar and spice.

Additionally, an equation for the marginal rate of substitution (MRS) is provided as an indicator showing the required relative proportion of sugar and spice:

$$MRS = \frac{\tau_1}{\tau_2}, \qquad (7.17)$$

where

$$\tau_1 = \frac{w_1}{m_1}, \ \tau_2 = \frac{w_2}{m_2}. \qquad (7.18)$$

A larger (smaller) MRS value indicates that the agent requires relatively more sugar (spice) than spice (sugar) to survive. Moreover, τ_1 and τ_2 correspond to the respective times until the agent dies of starvation if the amounts of procured sugar and spice fall below the necessary levels.

When spice is introduced into the experiment, agents located outside of the areas where the sugar and spice mountains overlap die in the initial iterations, since these agents can collect only one type of food. Lucky agents located in the region where the two mountains overlap procreate and gradually become capable of living in wider areas as their metabolic rates improve (Fig. 7.50).

Agents with high MRS (i.e., those that demand greater quantities of sugar than spice) are shown in a color close to cyan (light blue), agents with lower MRS (those that demand more spice than sugar) are shown in a color close to magenta (purple), and agents that have equal requirements for both foods

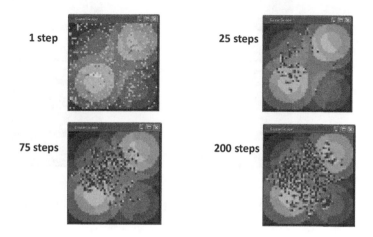

FIGURE 7.50: Agent aggregation with spice: (G_1, M).

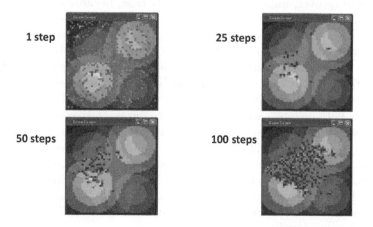

FIGURE 7.51: MRS values with spice: (G_1, M).

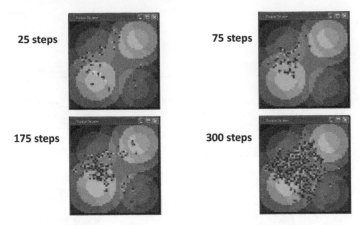

FIGURE 7.52: Agent aggregation with spice and combat.

are shown in a neutral color (Fig. 7.51). In the case here, agents located at the sugar mountains lack spice, and vice versa.

The average logarithmic *MRS* value for each agent stabilizes around 0 and shows that agents' overall collection of sugar and spice is well balanced.

Next, when the combat rule C_α from Section 7.10.6 is applied, the red and blue tribes fight over both sugar and spice. Although the average logarithmic *MRS* value for sugar and spice again stabilizes around 0, the blue tribe is annihilated, leaving only red agents (Fig. 7.52). The low-population tribe is unable to recover, unlike in the sugar-only case, because the remaining blue agents are located at the spice mountains and thus die due to a lack of sugar.

We now establish a rule for trading sugar and spice. Trade is conducted by adjacent agents, and the amounts traded are determined by the *MRS* values of both agents. Agents with sugar deficiency can obtain sugar by trading large quantities of spice, and vice versa. Furthermore, the direction of trade (i.e., which agent offers the sugar or spice) is determined by comparing the *MRS* values of both agents, and the agent that requires relatively more sugar trades spice, while the agent that requires more spice trades sugar. By conducting trade in this way, the difference between the *MRS* values of both agents decreases, and if the traded amounts exceed a certain value, the magnitudes of the agents' *MRS* values are reversed, indicating an exchange in the relative demands of sugar and spice. Therefore, further trade becomes counterproductive, and the ideal trading amounts are such that the *MRS* values of the two agents become equal.

Considering the above, the trading prices are determined as follows. When agents A and B exchange one unit of sugar for p units of spice, p is referred to as the trading price. Specifically, sugar is the product and p is the currency, where p is defined as the geometric mean of the *MRS* values of the two agents:

TABLE 7.12: Direction of trade [33].

Agent	MRS	
	$MRS_A > MRS_B$	$MRS_A < MRS_B$
A	Buying sugar and selling spice	Buying spice and selling sugar
B	Buying spice and selling sugar	Buying sugar and selling spice

$$p = \sqrt{MRS_A \cdot MRS_B}. \qquad (7.19)$$

Since the amounts of sugar and spice in Sugarscape are always integers, p is taken to be the closest integer value. Additionally, the direction of trade is determined in accordance with Table 7.12. The trading rule can be summarized as follows [33, p. 106]:

Trading rule T:

- Compare MRS values with an adjacent agent. Terminate the process if the values are equal, otherwise continue.

- The flow of spice is from the agent with high MRS toward the one with low MRS, and the flow of sugar is in the opposite direction.

- The geometric mean of the MRS values of the two agents is taken to be the trading price p.

- If $p > 1$, one unit of sugar is exchanged for p units of spice, and if $p < 1$, one unit of spice is exchanged for $1/p$ units of sugar.

- Repeat the trade until the step before the MRS values of the two agents are reversed.

The above trading rule is applied at each step for all adjacent agents.

In experimenting with this rule, scattering in the MRS values decreases when compared with the no trade case, and both sugar and spice are distributed in a well-balanced manner (Fig. 7.53). The traded amounts are about 50 units at each step, and the average trading price stabilizes at 1:1 approximately (Fig. 7.54). Since the respective amounts of sugar and spice growing in Sugarscape are approximately equal, their distribution becomes logarithmic. In addition, the metabolic rates of sugar and spice for all agents are set under identical conditions, and sugar and spice are expected to settle into an overall equivalence relation, demonstrating that an adequate distribution of wealth is achieved through skillful trade. Conversely, looking at the highest and lowest trading prices, cases exist where the values are different from 1 and do not converge even after 300 steps. Even if sugar and spice are used equivalently as a whole, local bias still exists, and trading continues. Such circumstances are easily overlooked in traditional economic analyses, which focus only on

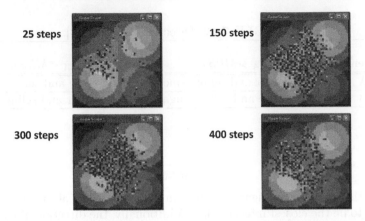

FIGURE 7.53: Agent aggregation with trade.

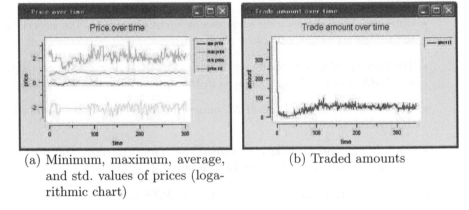

(a) Minimum, maximum, average, and std. values of prices (logarithmic chart)　　　(b) Traded amounts

FIGURE 7.54: Traded amounts and prices.

top-down balance. In other words, observing this type of behavior is one of the advantages of bottom-up multi-agent simulations.

Finally, macroscopic differences are introduced between sugar and spice. These differences are implemented by introducing alternating seasons, in which the growth rate of sugar is seasonally affected while that of spice is not. In this experiment, the seasons in the northern and southern hemispheres are swapped every 50 steps, and the growth rate for sugar in winter is set to 1/8.

Under these conditions, the average price oscillates over 50 step intervals (Fig.7.55). Comparing these results with the graph showing the average sugar reserves, it can be seen that the price of sugar increases as the sugar reserves decrease and clearly demonstrates the influence of environmental changes on trading prices. To elucidate the effects of trading on the agents, they are divided such that red agents are capable of trading and blue ones are not

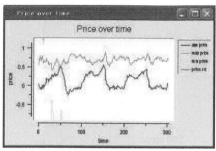
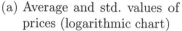

(a) Average and std. values of   (b) Average amounts of sugar
    prices (logarithmic chart)   and spice reserves

FIGURE 7.55: Trading prices and reserves in the case of seasonal change
and trade.

(Fig. 7.56). In this case, the red tribe becomes more populous, while the
population of the non-trading blue tribe decreases.

However, bias in the tribal populations is observed even when the experiment is conducted under the same conditions for both tribes. Hence, these
results alone do not show whether trade contributes to the superiority of the
red tribe. A further experiment is therefore conducted where trading capabilities are initially provided to the blue tribe and not to the red tribe, and
after 300 steps these conditions are reversed. The trading blue tribe becomes
consistently more populous until the 300th step, at which point the red tribe
gradually starts to trade and its population increases to that of the blue tribe.
Eventually, the red tribe became more populous than the blue tribe, demonstrating the role of trade in determining the superiority of a tribe.

7.10.8   Swarm simulation in Sugarscape

The main files of the system that implements "Sugarscape" are as follows:

`Sugarscape.java`	Source file of the main
`ObserverSwarm.java`	Source file of `ObserverSwarm`
`ModelSwarm.java`	Source file of the whole Swarm
`SugarAgent.java`	Source file of the agents
`SugarSpace.java`	Source file of the landscapes
`sugarspace.pgm`	File to place sugar
`spicespace.pgm`	File to place spices

The main parameters in `ModelSwarm` are shown in Table 7.13. Parameters
0 and 1 mean off and on, respectively. Do not forget to press "return" after
entering each parameter. Even after the execution starts, you can change
the parameters such as "color" if you stop to enter new values and press the
"start" button. The meaning of the GUI window of `ObserverSwarm` is given
in Table 7.14. "`displayFrequency`" and "`stopPeriod`" variables specify the

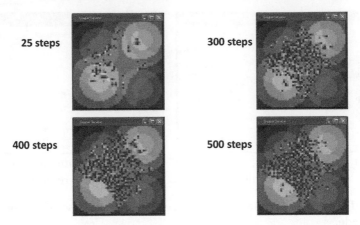

FIGURE 7.56: Agent aggregation when only red agents are capable of trading.

frequency of drawing and the interval of stopping, repectively. Figure 7.57 shows the state during the execution.

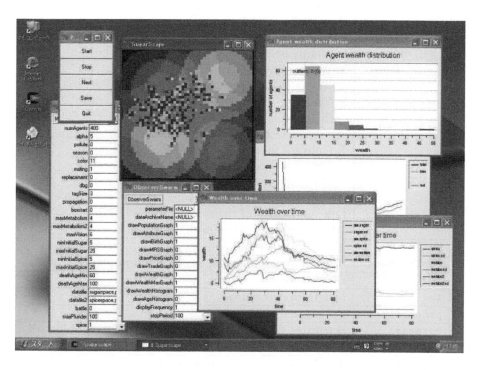

FIGURE 7.57 (See Color Insert): Sugarscape in Swarm.

TABLE 7.13: Parameters of Sugarscape.

Variable name	Meaning
numAgents	Number of initial agents
alpha	Foods' (sugar, spice) nutritive ratios (higher value, higher nutrition)
pollute	Generates pollution
season	Changes the seasons (1: sugar and spices, 2: only sugar, 3: only spices)
color	How to display agents: 0: Red color 1: Visibility (lower: blue higher: white) 2: Metabolic rate (lower: blue higher: white) 3: Wealth (little: blue bigger: white) 4: Age (smaller: blue higher: white) 5: Visibility (higher: aqua) and metabolic rate (higher: magenta) 6: Visibility (higher: aqua) and wealth (bigger: magenta) 7: Cultural tags (all 1: aqua all 0: magenta) 8: Cultural tags (more 1s: blue, more 0s: red) 9: Marginal rate of substitution *MRS* High: preference of sugar: aqua Low: preference of spice: magenta 10: Cultural tags (8 color display)
mating	Mates to have children
replacement	If one bug dies, one bug is generated randomly
dbg	Debug mode
tagSize	Length of cultural tag
propagation	Propagates a culture
boxstart	Separates the initial positions of red tribe and blue tribe.
datafile	The file name of sugar placement
datafile2	The file name of spice placement
battle	Fights with other tribes to rob sugar: 1: Careful 2: Without gun 3: Blue: without gun, Red: careful 4: Blue: careful, Red: without gun
maxPlunder	The maximum value of robbed sugar
spice	Feeds the spices after placement
trade	Trades sugar and spices (1: all agents, 2: only red tribe, 3: only blue tribe)

TABLE 7.14: GUI display of Sugarscape.

GUI name	Meaning
drawPopulationGraph	Graph of population (total, red tribe, blue tribe)
drawAttributeGraph	Graph (average, standard deviation) of visibility (average, standard deviation), metabolic rate (sugar, spice)
drawBirthGraph	Graph of number of births and deaths and number killed in battle
drawMRSGraph	Graph of marginal rate of substitution (average, maximum, minimum, standard deviation) of sugar and spice
drawPriceGraph	Graph of transaction costs (average, maximum, minimum, std dev)
drawTradeGraph	Graph of scale of trade
drawWealthGraph	Graph of wealth distribution
drawWealthMaxGraph	Graph of maximum value of wealth
drawWealthHistogram	Histogram of wealth distribution
drawAgeHistogram	Histogram of age distribution

Chapter 8

Conclusion

What is "real"? How do you define "real"? (from *The Matrix* movie quotes)

This book discussed the simulation of complex systems from the basics to implementation, applicable to a wide range of fields. In particular, implementation in Swarm, a simulation suite for complex systems, was explained in order to assist the reader to easily construct simulations. Complex systems and artificial life are actively being researched, which finds a place in many practical applications in the fields of animation and design, among others. Therefore, the objective of this book is to explain the fundamental concepts of artificial life and complex systems, and then to outline multi-agent simulation and bottom-up simulation principles.

The significance of bottom-up simulation of complex systems is analyzed in the summary of this chapter. The simulations presented in this book take a "constructive approach." This is an engineering approach that deepens understanding by building fundamental models and then observing the behavior of entire models built with those fundamental building blocks. The goal is to "create and understand" complex systems and artificial life.

The complex systems covered here are limited to actual systems in the real world because the objective of the constructive approach is to understand actual phenomena. Research is carried out sequentially in five stages. It gradually shifts from an abstract level to more concrete levels, and insights obtained in previous stages will be applied in subsequent stages. The research will go back to stage 1 after the completion of stage 5, and this procedure is repeated to increase understanding of the subject.

Stage 1 Make something, which may not necessarily correspond to the subject, that mimics the behavior of the subject. For example, models may be simulated as multi-agent simulations of complex systems such as in Swarm (Chapter 3).

Stage 2 Make something that qualitatively mimics the behavior of the subject. An example is the reproduction of ant marches (Section 5.1).

Stage 3 Make something that quantitatively mimics the behavior of the subject. Appropriate adjustment of the parameters of the simulator and comparisons with observable examples are necessary at this stage.

Stage 4 Hypotheses are tested, and the behavior of the model is adjusted to correspond to the actual behavior of the subject in the real world. Hypotheses are built to explain phenomena in natural science. Experiments are designed to verify whether the hypotheses hold. Similarly, the results of simulations are compared with the results of experiments in the real world as "hypothesis verification," aiming to connect the model to the real world.

Stage 5 Understand the origin of behavior in the subject, relate to the real world, and explain the cause and effect based on actual physical and chemical mechanisms. The goal is to understand not only the behavior of the subject but also to understand the underlying factors and the effects that the behavior causes. In biological terms, the objective is to understand proximate (physiological) and final (biological or evolutional) causes.

There has been much discussion on the guiding methodology on how to carry out research in the field of artificial life (see Section 4.4.1). For example, Barandiaran and Moreno categorized artificial life into following four models [9].

Generic model Aims to deduce generic properties that exist in any complex system.

Conceptual model Aims to formulate and understand concepts such as evolution or emergence.

Functional model Aims to understand specific systems with emergent properties.

Mechanistic model Aims to realistically reproduce the behavior of the subject model.

The relation between this classification of artificial life and the constructive approach is shown in Fig. 8.1. The complete correspondence to the subject in stage 5 is close to the mechanistic model. However, while the goal of a mechanistic model is to relate to the behavior of the subject, our approach aims to relate to factors that cause the behavior of the subject. Functional models model existing systems and correspond to stage 4 because the focus is mainly on emergence and evolution. In contrast, generic and conceptual models do not model actual subjects; therefore, strictly there is no one-to-one relationship with the stages in the constructive approach. However, these models can be considered as components of stages 2 to 4 because of the common objective of qualitatively and quantitatively clarifying the properties of complex systems.

Each model in artificial life research has a different viewpoint. On the other hand, the constructive approach progresses with research in steps with different viewpoints to understand existing phenomena. Deeper insight is obtained by connecting to the actual world, and then research is started from

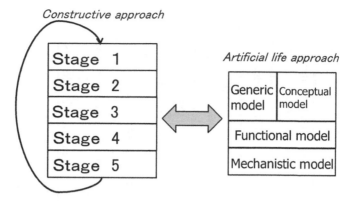

FIGURE 8.1: The classification of artificial life and the constructive approach.

stage 1 again using this insight. Repeating this approach will allow deeper understanding of the subject.

Tables 8.1 and 8.2 show the relations between related research and the constructive approach. Below we explain how this research corresponds to the stages.

Related research 1: Pattern formation

Alan Turing proposed that morphogenesis can be modeled by the reaction–diffusion of morphogens, which are virtual chemicals (see Section 7.7 for details). Genes corresponding to morphogens have been found recently, and research is under way. For instance, Yamaguchi et al. are investigating how patterns found on fish skin form at the genetic level based on Turing's reaction–diffusion model [129]. This work first built hypotheses based on the behavior of an actual model, and then simulations were carried out to finally derive the relationship to the model at the experimental biology level. Hence, this is an example where the constructive approach is taken.

Takeuchi et al. expanded the Turing model into a model that can explain the formation of bacteria and cancer cell colonies [112]. By controlling parameters, they verified that sufficient nutrition conditions would result in a spherical cancer tissue whereas deficient nutrition results in slow growth and an irregular, complicated colony.

Synthetic biology is a promising new field of genetic engineering in this direction [7, 111]. In synthetic biology, mathematical models are used to test biological hypotheses and observations, and to predict the possible behaviors of a designed gene circuit [49, 92]. These models serve as blueprints for novel synthetic biological systems, making the engineering of biology easier and more reliable. Evolutionary computation is a key tool in this field [58, ch. 6.5].

Related research 2: Artificial societies and artificial markets

TABLE 8.1: Relations between related research and the constructive approach (1).

	Stage 1	Stage 2	Stage 3	Stage 4	Stage 5
Pattern formation	Reproduction in a simulator [this book, Chapter 7]			Appropriate adjustment of parameters for pattern formation, hypothesis verification on reaction–diffusion model for patterns of fish [75], colony formation of skin cancer [61]	Genetic-level understanding of reaction–diffusion model for patterns of fish [75], understanding concrete criteria for choosing males and its stimulus process [129]. Synthetic biology approach [7, 49, 92, 111]
Artificial societies and artificial markets	Reproduction in a simulator [this book, Chapter 6]			Confirming similarities in wars and dealing [this book, Sections 6.5 and 6.7], verification of hypotheses such as the bandwagon effect [62]	Understanding emergence phenomena at the microscopic level (frequency distribution of rate fluctuation, contrary opinion phenomena) through interviews with dealers [62]
Acoustic consonance and dissonance perception model	Theory of beats (Helmholtz), dissonance perception model (Kameoka and Kuriyagawa) [63, 64]	Investigating similarity to consonance and dissonance perceived by humans [63, 64]	Agreement with psychological experiments [63, 64]	Prediction and verification of consonance in arbitrary sound [63, 64]	Frequency perception mechanism of the inner ear and relation to model

TABLE 8.2: Relations between related research and the constructive approach (2).

	Stage 1	Stage 2	Stage 3	Stage 4	Stage 5
Biological speciation	Reproduction in a simulator [17, 84]		Investigating the number of species that will emerge [17, 84]	Investigation of how mutation influences speciation [84]	Agreement with experiments on bacteria [84]
Foraging of animals	Simulation using a classifier system [56]		Acquisition of optimized foraging rules [56]	Foraging based on a numerical model [26], mimicry and search [94]	Experiment to verify model on cognition during foraging [36]
Evolutionary robotics	Evolving virtual creatures [107], building robots that can perform specific tasks [11, 27, 39]		GOLEM project (NASA) [96], morphogenesis of robots [117], building robots capable of autonomous learning [11, 27, 39]	Experiment with walking humanoids with brains mimicking monkeys [68], experiment with robots that control insects [113], building robots that can perform specific tasks [11, 27, 39]	
Emergence in army ants	Reproduction in a simulator [this book, army ant]	Reproduction of appropriate shortcuts [this book, army ant]	Observation of collective determination [this book, Section 2.5]	Hypothesis verification of collective determination [this book, Section 2.6]	Connection with actual army ants

Sugarscape, which was proposed by Joshua Epstein and Robert Axtell [32], is a model that forms an artificial society (see Section 7.10 for details). This is basically a simulation where ants move around looking for sugar, but its objective is to understand mechanisms behind social behavior by taking into account concepts such as mating, warfare, and dealing. Understanding the environments around each agent and under what conditions agents engage in warfare or make deals would help us understand the behavior of actual economies. For example, Kiyoshi Izumi [62] attempted to explain macroscopic phenomena such as the frequency distribution of rate fluctuation or contrary opinion by laws of cause and effect at the microscopic level through interviews with dealers (see [59]). This corresponds to stage 5 research that aims to connect emergent phenomena to the actual world.

Related research 3: Acoustic consonance and dissonance perception model

Humans perceive two or more combined sounds as a pleasant consonance or an unpleasant dissonance. Perception models of consonance and dissonance are based on Helmholtz's theory of beats where the extent of consonance depends on the closeness of harmonics [48]. This theory was formulated mathematically by Kameoka and Kuriyagawa [63, 64], and its internal parameters were determined by psychological experiments. The mechanism of the inner ear that picks up frequency, which corresponds to the input in the model, has been clarified; however, the relation to the internal mechanism of the brain is still not understood. This connection would correspond to stage 5.

Related research 4: Biological speciation

Clement researched biological speciation through an investigation of the ecology of fish [17]. Artificial life that simulated fish was created to research what kind of clustering is effective in speciation. Metivier et al. investigated how stress affects speciation of individuals through a simulator called Life Drop [84]. The simulator developed showed that stress from the environment increases the possibility of crossover between species, which in turn affects speciation. The conclusion was that stress strongly influences evolution. The simulation results agreed with the results of biological experiments using bacteria, and hence this corresponds to stage 5 in the constructive approach.

Related research 5: Foraging of animals

There has been much research on optimization of the foraging strategy of animals in mathematical ecology [61] and behavioral ecology [26]. For instance, assume the nutrition values (g_i) and the cost necessary for intake (h_i) are given for multiple types of food. The hypotheses derived from the theory to optimize foraging are the following.

Claim 1 If every type of food exists with the same distribution, the food with higher $\frac{g}{h}$ has higher preference.

Claim 2 If there is an abundance of a preferred food (food with large $\frac{g}{h}$), only that food should be eaten.

Claim 3 The decision to eat or not to eat a preferred food does not depend on the amount of less-preferred foods.

The author used simulations, for instance to learn optimization strategies using classifier systems, and confirmed that rules that are consistent with these hypotheses could be obtained [56]. There is also an investigation on the searching pattern that results in optimized foraging (cognitive model for efficient foraging). Erichsen et al. experimentally verified cognitive models using paridae [36].

Related research 6: Evolutionary robotics

Karl Sims [107] created evolved virtual creatures, or creatures made of directional graphs that have actions to mimic creatures (see Section 4.4.2 for details). These creatures may show shapes and actions that people cannot even conceive of, and one study applied these creatures to morphogenesis of robots [117]. A project, RobotCub, aims to create robots with the learning ability of human toddlers through a constructive approach [11, 27, 39]. The objective of this research is to clarify how toddlers learn specific tasks and what is necessary in the learning process. Mitsuo Kawato created walking robots that reproduce information on brain activity in walking monkeys to gain a deeper understanding of the brain [68]. Creating actual models to understand actual subjects and to make connections corresponds to stage 5. Takashima et al. carried out research to understand the mechanisms in the brain by connecting silkworms or their brains to robots to make silkworms intelligence control robots [113].

Related research 7: Emergence in army ants

The research introduced in Section 5.8 aims to reproduce and understand the cooperative behavior of army ants. Experiments using simulators designed to reproduce cooperative behavior in army ants showed that agents built bridges in appropriate places to make shortcuts. This is an accomplishment of stage 2. Incorporating different collaborative behavior and hypotheses considering the ecology of ants in the simulator resulted in agents initially attempting to build bridges in various positions and later concentrating on building existing bridges located in various positions. This is similar to the collective decision-making of ants in nature, and corresponds to stage 3. However, this simulation does not connect directly to the actual world; therefore, the current goal is to repeat hypothesis verification to connect more closely to the results of experiments involving animals (stage 4). The environments in a simulator in which agents move around are currently limited to simple environments. Connection to the actual world will be attempted by adjusting factors such as the field in which the agents move, the size of the agents, or the variation in velocity of agents to make the environment in the simulator similar to the real environment (examples include the size of the experimental equipment, number of ants, and speed of ants [80, 81]) in experiments involving animals.

In this book, both the constructive approach and actual examples have been introduced to help clarify complex systems and artificial life. Past research studies in this field show that the investigation of cause and effect is of paramount importance. Therefore, current research has been presented as actual examples for showing how to perform research using the bottom-up approach. Various multi-agent simulation tests implemented in this book have been categorized and discussed in relevant sections.

The approach explained in this book may not be applicable for research studies dealing with some types of complex systems. However, structural models that can be imagined and constructed are necessary for simulations in complex systems science, and the models become closer to reality by "identification of simulations." The multi-agent simulations discussed in this book will increase the understanding of target phenomena when actual models exist. This approach aims to investigate the behavior of actual targets, i.e., the cause and effect phenomenon, and then proceed to researching the systems in a step-by-step manner. This book would immensely contribute to research in the fields of complex systems and artificial life.

Appendix A

GUI Systems and Source Code

A.1 Introduction

For the sake of better understanding GP and its extensions, software described in this book can be downloaded from the website of the author's laboratory (http://www.iba.t.u-tokyo.ac.jp/). They are LGPC for Art, TSP by GA, GP, PSO, etc.

The intention of making these software packages available is to allow users to "play" with emergent systems. We request interested readers download these packages and experience the emergent properties. The web pages contain (1) instructions for downloading the packages, (2) a user's manual, and (3) a bulletin board for posting questions. Readers are invited to use these facilities.

Users who download and use these programs should bear the following items in mind:

(1) We accept no responsibility for any damage that may be caused by downloading these programs.

(2) We accept no responsibility for any damage that may be caused by running these programs or by the results of executing such programs.

(3) We own the copyright to these programs.

(4) We reserve the right to alter, add to, or remove these programs without notice.

In order to download the software mentioned in this book, please follow the following link, which contains further instructions:

http://www.iba.t.u-tokyo.ac.jp/english/BeforeDownload.htm

If inappropriate values are used for the parameters (for instance, elite size greater than population size), the program may crash. Please report any bugs found to stroganoff@iba.t.u-tokyo.ac.jp.

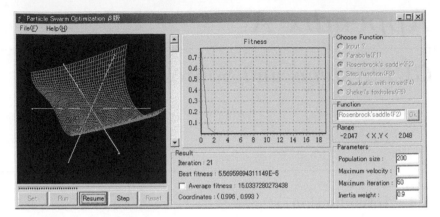

FIGURE A.1: PSO simulator.

A.2   PSO simulator and benchmark functions

The PSO search process could be observed using this simulator (see Fig. A.1). The user is able to freely define the functions for the search.

In addition, this simulator uses De Jong's standard functions. De Jong's standard functions are benchmark tests for GAs, and are used for determining the minimum value. The definitions of these functions, along with the definition fields and optimum values, are shown in Table A.1. The form of the functions and plots projected on the $x_1 - x_2$ plane are shown in Fig. A.2. Benchmark functions $F4$ and $F5$ seem to be more difficult than the others. The $+GAUSS(0, 1)$ of $F4$ shows the addition of values from the normal distribution with average 0 and dispersion 1. In other words, noise is included in the various points in $F4$. With $F5$, there is a series of 5×5 valleys lined up in a grid alignment, but the valleys do not have a uniform depth. The trough of the valley at the lowermost left is the minimum value (≈ 1), while the local minimum values of the remaining troughs increase sequentially from left to right and from bottom to top, as 2, 3, etc. When leaving these troughs, it rapidly approaches the maximum value of 500. Note that the coordinates of a_{ij} are as follows:

```
int a[2][25] = {
    {-32, -16, 0, 16, 32, -32, -16, 0, 16, 32, -32, -16, 0, 16, 32,
     -32, -16, 0, 16, 32, -32, -16, 0 16, 32},
    {-32, -32, -32, -32, -32, -16, -16, -16, -16, -16, 0, 0, 0, 0,
     0, 16, 16, 16, 16, 16, 32, 32, 32, 32, 32}
};
```

Originally, $F1$, $F2$, and $F3$ can have generalized definitions with three or

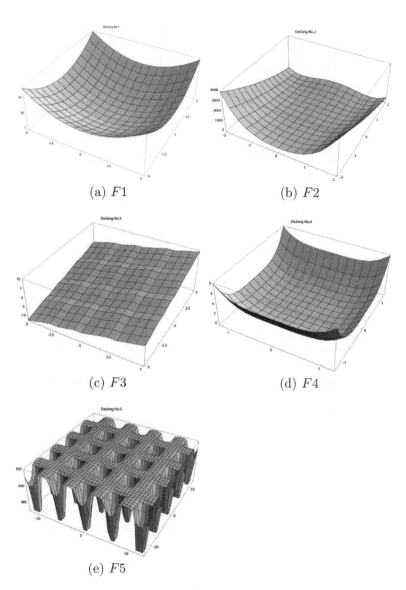

(a) $F1$

(b) $F2$

(c) $F3$

(d) $F4$

(e) $F5$

FIGURE A.2: Benchmark functions.

TABLE A.1: De Jong's standard functions.

Function Name	Definition	Domain	Optimum Value
$F1$	$\sum_{i=1}^{3} x_i^2$ Paraboloidal surface	$-5.11 \le x_i < 5.12$	0
$F2$	$100(x_1^2 - x_2)^2 + (1 - x_1)^2$ Rosenbrock's saddle	$-2.047 \le x_i < 2.048$	0
$F3$	$\sum_{i=1}^{5} \lfloor x_i \rfloor$ Step function	$-5.11 \le x_i < 5.12$	-30
$F4$	$\sum_{i=1}^{30} ix_i^4 + \mathrm{GAUSS}(0,1)$ Quartic function with noise	$-1.27 \le x_i < 1.28$	0
$F5$	$\left[\frac{1}{500} + \sum_{j=1}^{25} \frac{1}{j+\sum_{i=1}^{2}(x_i - aij)^6} \right]^{-1}$ Shekel's foxholes	$-65.535 \le x_i < 65.536$	1

more variables, but here we reduced the number of dimensions in the functions to make two-variable problems that are much simpler to work with.

For convenience, the Z axis was greatly compressed in the view of $F5$ in the PSO simulation, so it looks somewhat different from the previous figure.

Download and unzip the file "Particle Swarm Optimization ver1.0" (402 kB). It contains 3 files: `EquToDbl.dll`, `EquToDbl.txt`, and `PSO.exe`. Click `PSO.exe` to start the simulator.

The following commands allow the user to make basic use of the simulator:

- Set button
  When pressed, an initial population is generated.

- Run button
  When pressed, execution is initiated.

- Stop button
  Pressing this button halts calculations. It is used when the user wants to observe the movement of individuals during the simulation.

- Step button
  This button can be used after pressing the Stop button to sequentially observe motions at each generation.

- Reset button
  After a simulation has been completed, pressing this will re-start execution.

The following parameters can be set by the user.

- Population size
  Number of individuals in the population.

- Maximum velocity
  The maximum velocity of any individual in motion. All individuals are prevented from moving any faster.

- Maximum iteration
  Maximum number of replications of the simulation.

- Inertia weight
  Attenuation coefficient. The default value is 0.9, causing the speed to gradually decrease with time.

- Input?
  If this box is checked, the user is able to freely define the functions.

After execution is started, the fitness transition is plotted in the chart area at the center of the screen. The following items are displayed in the "Result" panel beneath the plot:

- Iteration
  Number of the current replication. This corresponds to the number of generations in a GA.

- Best fitness
  Best (minimum) fitness value among all individuals.

- Average fitness
  Average fitness value of all individuals. If this box is checked, the average fitness is plotted in the above graph.

- Coordinates
  This shows the coordinates of the best fitness.

A.3 TSP simulator by a GA

The TSP simulator works as follows.

(1) Set values for number of city, population size, max generation in the text boxes next to the labels City, Population Size and Max Generation, respectively.

(2) Also set crossover and mutation rates. Select "selection method" and "replacement strategy" by clicking the choice provided. If you choose elitism replacement, provide the percentage value for elite replacement. If none is changed, default values are used.

(3) Either click the Random Point button or click on the drawing area and put city locations one by one.

(4) Click the Run button.

Parameters and execution conditions can be changed through the following procedure during runtime.

- When in Running mode, you can change the Selection Method, Replacement Strategy, Delay Rate, Crossover Rate, Mutation Rate, and Elitism.

- When in Running mode, you can change the city location by clicking the city point and dragging it to a new location.

You can just click the Stop button to stop running. You can rerun the program by just clicking the Run button after clicking the Stop button or after normal stopping. This can be done only when the Run button is enabled. When the Reset button is enabled, you may click it and put the desired values in the corresponding text boxes.

When in running mode, RED LINES and YELLOW LINES show the best-ever tour found and the best tour in the current generation, respectively.

The best-ever tour and the best tour in the current generation are displayed along with their distances in the textboxes next to the labels: Overall Best, Current Generation, and Distance, respectively. The Current Generation number is displayed in the textbox labeled Generation#.

A.4   Wall-following simulator by GP

A robot learns through genetic programming (GP) in this code such that it is programmed to move along a wall in a room with obstacles (Fig. 2.12). A `jar` file is released, and a Java runtime environment is necessary to execute this code.

Each individual (robot) is shown by a blue circle (the front side of the robot is indicated with a line). The size indicates the fitness (Fig. 2.15), where the fitness is the number of tiles adjacent to a wall that the robot passed, and larger size means better fitness.

The robots evolve such that they follow walls better as the number of generations increases. The robots become larger, showing that performance is improving. The behavior of all individuals in the population is displayed simultaneously, but there is no cooperation between them.

The three lines at the bottom show the following information.

- Best Fitness: the fitness of the best individual in this generation

- Best tree: the S-expression (a notation for tree-structured data) of the best individual in this generation

---

A.5   CG synthesis of plants based on L-systems

This system draws trees based on an L-system using interactive evolutionary computation (IEC) (Fig. 2.17).

This can be executed by entering

```
java -jar treeIEC.jar
```

from a terminal, or by double-clicking `treeIEC.jar` in Windows.

When you pick two trees from eight trees and click a button, eight new trees are generated that have the characteristics of the two trees. Repeating this procedure by continuing to click on two trees generates a tree that you like. The same tree can be clicked twice to generate next-generation trees that reflect the characteristics of this tree only.

A data file with the name **data** will be saved in the folder when the **save** button is clicked. Tree data can be loaded from the data file when the **load** button is clicked.

A.6   LGPC for art

This simulator designs abstract figures (wallpaper) based on IEC methods (Fig. 2.18). The basic procedure for using this simulator is as follows.

(1) Click Clear if you do not like the pictures shown in the 20 windows in the View tab; all windows will be initialized.

(2) If you like any of the pictures, select by clicking on it (its frame becomes red). Any number of pictures can be selected.

(3) When you click OK, the selected pictures are used as parent candidates to generate and show a population of next-generation genes.

(4) Repeat (1) to (3)ĄD

You can save genes that you like (Gene_Save command), or load a picture that you had previously saved and replace a displayed picture with it (Gene_Load command). The 20 windows in the View tab are numbered to allow replacement (the top-left picture is No. 1; the number increases from left to right, top to bottom up to No. 20).

The following parameter settings and items are shown on the GP Parameters screen.

- coordinate
  This determines the origin of the coordinates in showing the expressions. Selecting the upper-left results in an origin at the top-left corner of the frame, whereas selecting the center results in an origin at the center (this makes the pattern more likely to make the top and bottom halves and/or the right and left halves symmetric).

- OK
  This is possible when at least one figure is selected. Clicking this button results in derivation of a new generation where genes of selected pictures are parent candidates, and a new generation is generated and displayed.

- Clear
  This is possible when no figure is selected. Clicking this button initializes the gene population.

The other parameters and status are as follows.

- Results screen (Fig. A.3(a)): Settings and display of GTYPE

  - Functions
    This determines the functions used as non-terminal symbols in GP. Functions to be used should be checked.

  - Constants
    This determines the constants used as terminal symbols in GP. The range and interval to be used should be entered.

  - Best Individual
    The GTYPE of the best individual (in case of LGPC for Art, all individuals) in the generation is displayed.

- GP Parameters screen (Fig. A.3(b)): The problem is defined.

  - Number of Populations (not available in LGPC for Art)
    This determines the number of populations. Each population can evolve using different parameters.

  - Population Size (not available in LGPC for Art)
    This determines the number of individuals.

  - Generation (not available in LGPC for Art)
    This determines the number of generations.

  - Selection Method (not available in LGPC for Art)
    The selection strategy is determined from proportional, tournament, or random. Whether to implement an additional elitist strategy is also determined.

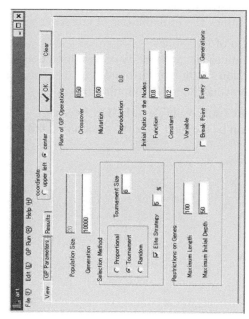

(a) Results screen

(b) GP Parameters screen

FIGURE A.3: LGPC for art.

– Restriction on Genes
  This determines the largest size of a gene.

– Rate of GP Operations
  This determines the probabilities of crossover and mutation.

– Initial Ratio of the Nodes
  This determines the probability of assigning nodes in each individual as function symbols or terminal symbols (constants or variables) when generating the initial population.

– Break Point (not available in LGPC for Art)
  Searches can be stopped after the designated number of generations elapses.

Appendix B

Installing Swarm

In this section, we explain how to install the Java version of Swarm 2.2 on a Windows machine (as of August 2012). Users can search the web or visit the author's home page[1] for the latest information related to installing Swarm on other operating systems or on other Windows versions.

B.1 Installation

The following components are required to use Swarm:

- the Java SDK
- the Swarm package
- the modified scripts

These are described below.

B.1.1 Java2 SDK installation

Download the JS2E SDK (approximately 52 MB) from the Java home page.[2] J2EE can also be used. Click "Accept" when asked if you accept the license terms, then "Continue." Next, download the following file:

```
Windows Offline Installation, Multi-language
  j2sdk-1_4_2_11-windows-i586-p.exe
```

Start the installer, and answer the questions that are displayed.

It is also possible to use the latest version J2SE (Standard Edition 5.0, `jdk-1\_5\_0\_07-windows-i586-p.exe`, 63.43 MB), but see the author's web site[3] for information related to compatibility with Swarm 2.2.

---

[1]http://www.iba.t.u-tokyo.ac.jp/~yanase/swarmlecwiki/?Install
[2]http://java.sun.com/j2se/1.4.2/ja/download.html
[3]http://www.iba.t.u-tokyo.ac.jp/~yanase/swarmlecwiki/?Install

B.1.2   Installing the Swarm package

Download the latest version of the Windows binary Swarm package from the stable version download page[4] (`Swarm-2.2-java.zip` is approximately 6.3 MB).

Unzip the file to `C:\` (or to some other location) and mount it. After unzipping the file, you should have a directory like `C:\Swarm-2.2-java`.

Next, create the source folder (which is where the program that we are about to create will be stored). Below, we will use the example `C:\Swarm-2.2-java\home`.

B.1.3   Setting environment variables

Open a command prompt and go to the source directory.

```
cd C:\Swarm-2.2-java\home
```

Next, enter the following commands at the prompt:

```
set PATH=%PATH%;C:\j2sdk1.4.2_11\bin
set PATH=%PATH%;C:\Swarm-2.2-java\bin
set CLASSPATH=%CLASSPATH%;C:/Swarm-2.2-java/share/swarm/swarm.jar
set CLASSPATH=%CLASSPATH%;C:/Swarm-2.2-java/share/swarm/kawa.jar
set SWARMHOME=C:/Swarm-2.2-java
```

Note that the path separator for SWARMHOME must be a "/" character. The CLASSPATH setting can also use "\" characters.

It is also possible to set the environment variables by copying the above into a batch file (e.g., `setup.bat`) and executing it. Another possibility is adding the above environment variables to your system environment so that the above process does not have to be performed with each launch.

B.1.4   Compiling

To compile the program, execute the following in the directory with the java file:

```
javac *.java
```

To execute the program, type

```
java "class file containing the main() function"
```

For example,

```
java StartHeatbugs
```

---

[4]http://www.swarm.org/wiki/Swarm:_stable_release

B.1.5 Confirming operation

The Santa Fe Institute distributes two sample files, `jheatbugs` and `jmousetrap`. Below we will demonstrate the compilation and execution of `jheatbugs`.

Download `jheatbugs-2.1.tar.gz` from within java→sdg on the Swarm official website:

`http://ftp.swarm.org/pub/swarm/apps/`

After downloading, unzip it and run the following command:

```
javac *.java
java StartHeatbugs
```

If the build fails, or when you wish to compile again, please recompile after running the following command to delete all class files:

```
del *.class
```

You can also execute `jmousetrap-2.1.tar.gz` in a manner similar to the above.

B.2 Objective-C version

This book describes the Java version of Swarm, but there is also an Objective-C version of the program. The following is a supplemental description of that version.

B.2.1 Objective-C and Swarm

The Objective-C programming language is an extension of C for object-oriented programming. The language is based on Smalltalk and is currently the standard language for Mac OS X programming. On Linux and other Unix-compatible systems, the GNU GCC compiler also supports Objective-C through a suite of libraries called GNUStep.

Swarm was originally developed in Objective-C, and while a Java version exists, the Objective-C version remains the primary development target (making the Java version something of a side project). In the Java version, all class design and method naming rules follow Objective-C standards. This in turn means that Java conventions are often not followed, which can be somewhat confusing to a user who is used to Java norms.

In terms of implementation, the Java version of Swarm is somewhat unstable. The Windows edition of the Java version has also been reported to be slow.

When using the Java version of Swarm, should you feel that the simulation speed is too slow to be usable, or that bugs are preventing desired operation, you can consider switching to the Objective-C version.

B.2.2 Material related to the Objective-C version of Swarm

The Swarm tutorial described in Chapter 3 is actually an adaptation of documents originally written for the Objective-C version by Dr. Benedikt Stefansson, rewritten for Java by our staff members.

Differences in the languages have resulted in some differences in the code, but the description should be sufficient for understanding the Objective-C version.

Note that there is also an Objective-C programming guide on the main Swarm site.[5]

B.2.3 Running Swarm under various environments

It is possible to use the Objective-C version of Swarm on all platforms that Swarm supports. In fact, under Mac OS X the Objective-C version is the only one available. Below are some notes related to running the Objective-C version under various operating systems.

B.2.3.1 Windows

When running the Objective-C version of Swarm under Windows, it is necessary to use a version of the Unix compatibility environment Cygwin[6] specially prepared for Swarm. It is difficult to simultaneously use both the Swarm version of Cygwin and the normal Cygwin, so if you normally use Cygwin for other purposes please consider using VMWare Player,[7] coLinux,[8] or some other free virtual machine environment to create a Linux guest environment and run Swarm within it.

B.2.3.2 Unix

The Objective-C version of Swarm should run bug-free under Solaris, IRIX, HP-UX, and other flavors of Unix, as well as PC-Unix and, in particular, Linux.

The current main target for Swarm development is Linux, and the various distribution binaries are available on many websites. RPMs for Red Hat-based Linux distributions such as Fedora are available from the Swarm site. There is also a package for Debian.

For other Linux distributions, as well as PC-Unix and Unix, it is necessary

---

[5]http://www.swarm.org/swarmdocs-2.2/set/set.html
[6]http://cygwin.com/
[7]http://www.vmware.com/products/player/
[8]http://www.colinux.org/

to compile the program from source and install manually.[9] Note, however, that the source code packages use autoconf, so there should be no need to edit makefiles and the like, making installation relatively painless.

There might be some rare cases where the default GCC compiler on some Linux and PC-Unix distributions does not support Objective-C. This will prevent successful compilation of the Swarm libraries, but this problem can be remedied by installing a version of GCC that supports Objective-C.

B.2.3.3 Mac OS X

It is not currently possible to use the Java version of Swarm under Mac OS X, making the Objective-C version the only option.

Under Mac OS X, Swarm runs under X11.app, Apple's X-Window environment. This means that it is necessary to install X11.app when installing the operating system.

Packages containing binaries for OS X are distributed on the main Swarm site.[10] Simply downloading and installing these binaries should allow the creation of a Swarm environment.

B.3 Useful online resources

There is a wealth of information about Swarm available on the Internet. Below is a list of just some. We have omitted some resources that are very easily found by a web search.

SwarmWiki Developer site.[11] The latest information about Swarm should be available here.

Swarm Development Group The main site before creation of the Wiki. The tutorials here[12] are an excellent place to start (but note that most use Objective-C).

Swarm On-Line FAQ The first place to visit when you have a question.[13] There are also many usage cases here.

Swarm API reference A reference for the Swarm API.[14] This will be absolutely necessary if you intend to write your own Swarm applications. You will also want to keep the Java API reference[15] handy.

---

[9]http://www.swarm.org/wiki/Swarm:_platforms
[10]http://www.swarm.org/wiki/Swarm:_MacOS_X_binaries
[11]http://www.swarm.org/wiki/Main_Page
[12]http://www.swarm.org/intro-tutorial.html
[13]http://www.ku.edu/~pauljohn/SwarmFaq/
[14]http://www.santafe.edu/projects/swarm/swarmdocs/refbook-java/index.html
[15]http://java.sun.com/j2se/1.3/ja/docs/ja/api/index.html

SwarmFest The international Swarm conference. Joining will allow you to view many examples of research done using Swarm.

Another excellent reference is Dr. Taro Yabuki's development notes page and collection of help answers.[16]

---

[16]http://www.iba.t.u-tokyo.ac.jp/~yabuki/tip/swarm/swarm.html

References

[1] Anderson, C., Theraulaz, G. and Deneubourg, J.L.: "Self-assemblages in insect Societies," *Insectes sociaux*, vol. 49, no.2, pp. 99–110, 2002.

[2] Andre, D., Bennett III, F.H., and Koza, J.: "Evolution of intricate long-distance communication signals in cellular automata using genetic programming," in *Artificial Life V: Proceedings of the Fifth International Workshop on the Synthesis and Simulation of Living Systems*, 1996.

[3] Angeline, P.J.: "Evolutionary optimization versus particle swarm optimization: Philosophy and Performance differences," *Evolutionary Programming VII*, Porto, V.W., Saravanan, N., Waagen, D., and Eiben, A.E. (eds.), pp. 601–610, Berlin, Springer, 1998.

[4] Angtuaco, S.P.: "Amazing Ants: How to Form a Bridge," http://jill-of-alltrades.hubpages.com/hub/Amazing-Ants.

[5] Axelrod, R.: *The Evolution of Cooperation*, Basic Books, New York, 1984.

[6] Axelrod, R.: "An evolutionary approach to norms," *American Political Science Review*, vol. 80, no.4, pp. 1095–1111, 1986.

[7] Basu, S., Gerchman, Y., Collins, C.H., Arnold, F.H., and Weiss, R.: "A synthetic multicellular system for programmed pattern formation," *Nature*, vol. 434, pp. 1130–1134, 2005.

[8] Ando, D. and Iba, H.: "Real-time musical interaction between musician and multi-agent system," *Proceedings of the 8th International Conference on Generative Art 2005*, Milan, Italy, pp. 93–100, 2005.

[9] Barandiaran, X. and Moreno, A.: "ALife models as epistemic artefacts," *Artificial Life X: Proceedings of the Tenth International Conference on the Simulation and Synthesis of Living Systems*, pp. 513–519, 2006.

[10] Bentley, P.J.: *Evolutionary Design by Computers*, Morgan Kaufmann, San Francisco, CA, 1999.

[11] Berthouze, L. and Metta, G.: "Epigenetic robotics: modelling cognitive development in robotic systems," *Cognitive Systems Research*, vol. 6, Issue 3, pp. 189–192, 2008.

[12] Biles., J.A.: "Life with GenJam: Interacting with a musical IGA," *Proceedings of IEEE International Conference on Systems, Man, and Cybernetics*, pp. 652–656, 1999.

[13] Brackman, A.C.: *A Delicate Arrangement: The Strange Case of Charles Darwin and Alfred Russel Wallace*, Times Books, New York, 1980.

[14] Caldwell, C. and Johnston., V.S.: "Tracking a criminal suspect through "face-space" with a genetic algorithm," *Proceedings of the Fourth International Conference on Genetic Algorithms (ICGA91)*, pp. 416–421, 1991.

[15] Caro, G. and Dorigo, M.: "AntNet: A mobile agents approach to adaptive routing," Tech. Rep. IRIDIA/97-12, Universite Libre de Bruxelles, Belgium, 1997.

[16] Cicirello, V. and Smith, S.: "Improved routing wasps for distributed factory control," *Proceedings of IJCAI-01 Workshop on Artificial Intelligence and Manufacturing: New AI Paradigms for Manufacturing*, pp. 26–32, 2001.

[17] Clement, R.: "Visualising speciation in models of cichlid fish," *Proceedings of the 17th European Simulation Multiconference*, pp. 344–348, 2003.

[18] Clerc, M. and Kennedy, J.: "The particle swarm: Explosion, stability, and convergence in a multidimentional complex space," *IEEE Transactions on Evolutionary Computation*, vol. 6, no.1, pp. 58–73, 2002.

[19] Cohen, M.D., Riolo, R.L., and Axelrod, R.: "The emergence of social organization in the prisoner's dilemma: how context-preservation and other factors promote cooperation," Santa Fe Institute Working Paper 99-01-002, URL: http://cscs.umich.edu/research/techReports.html, 1999.

[20] Collins, R.J., and Jefferson, D.R.: "The evolution of sexual selection and female choice," *Proceedings of the First European Conference on Artificial Life (ECAL92)*, pp. 327–336, MIT Press, Cambridge, MA, 1992.

[21] Cronin, H.: *The Ant and the Peacock: Altruism and Sexual Selection from Darwin to Today*, Cambridge University Press, New York, 1993.

[22] Darwin, C., *On the Origin of Species by Means of Natural Selection, or the Preservation of Favoured Races in the Struggle for Life*, John Murray, London, 1859.

[23] Darwin, C.: *The Descent of Man*, D. Appleton and Company, New York, 1871.

[24] Dawkins, R.: *The Blind Watchmaker*, W.W. Norton, New York, 1986.

[25] Dawkins, R.: *The Selfish Gene*, Oxford University Press, Oxford, UK, 1991.

[26] Krebs, J.R. and Davies, N.B.: *An Introduction to Behavioural Ecology*, Wiley-Blackwell, Oxford, UK, 1993.

[27] Degallier, S., Righetti, L., Natale, L., Nori, N., Metta, G. and Ijspeert, A.: "A modular, bio-inspired architecture for movement generation for the infant-like robot iCub," *Proceedings of IEEE RAS/EMBS International Conference on Biomedical Robotics and Biomechatronics (BioRob2008)*, 2008.

[28] Deneubourg, J.L., Gross, S., and Franks, N.R.: Sendova-Franks, A., Detrain, C., and Chretien, L.: "The dynamics of collective sorting: Robot-like ants and ant-like robots," *Proceedings of Simulation of Adaptive Behavior: From Animals to Animats (SAB91)*, pp. 356–363, 1991.

[29] Dewdney, A.K.: "Simulated evolution: Wherein bugs learn to hunt bacteria," *Scientific American,* pp. 138–141, 1989.

[30] Dirk, M.: Max-Planck-Institut fur Radioastronomie, http://www.mpifr-bonn.mpg.de/staff/dmuders/

[31] Dorigo, M. and Gambardella, L.M.: "Ant colonies for the traveling salesman problem", Tech. Rep. IRIDIA/97-12, Universite Libre de Bruxelles, Belgium, 1997.

[32] Epstein, J.M. and Axtell, R.: *Growing Artificial Societies: Social Science from the Bottom Up,* A Bradford Book, 1996.

[33] Epstein, J.M. and Axtell, R.: *Growing Artificial Societies,* MIT Press, 1996.

[34] Eberhart, R.C. and Hu, X.: "Human tremor analysis using particle swarm optimization," *Proceedings of IEEE Congress on Evolutionary Computeration 1999,* pp. 1927–1930, 1999.

[35] Eberhart, R.C. and Shi, Y.: "Comparison between genetic algorithms and particle swarm optimization," *Proceedings of the Seventh Annual Conference on Evolutionary Programming,* pp. 611–619, 1998.

[36] Erichsen, J.T., Krebs, J.R., and Houston, A.I.: "Optimal foraging and cryptic prey," *Journal of Animal Ecology,* vol. 49, pp. 271–276, 1980.

[37] Fabre, J.-H.: *Insect Adventures Jean-Henri Fabre,* Alexander Teixeira De Mattos (Translator), Kessinger Publishing, Whitefish, Montana, 2005.

[38] Fisher, R.A., The Genetical Theory of Natural Selection, Clarendon Press, New York, 1930.

[39] Fitzpatrick, P., Needham, A., Natale, L., and Metta, G.: "Shared challenges in object perception for robots and infants," *Infant and Child Development,* vol. 17, Issue 1, pp. 17–24, 2008.

[40] Frisch, U., B. Hasslacher, B., and Pomeau, Y., Lattice-gas automata for the Navier-Stokes equation. *Physical Review Letters*, vol. 56, pp. 1505–1508, 1986.

[41] Gacs, P., Kurdyumov, G. L., and Levin, L. A., "One-dimensional uniform arrays that wash out finite islands," *Problemy Peredachl Informatsii*, vol. 12, pp. 92–98, 1978.

[42] Gaing, Z.-L.: "Particle swarm optimization to solving the economic dispatch considering the generator constraints," *IEEE Transactions on Power Systems*, vol. 18, no.3, pp. 1187–1195, 2003.

[43] Goss, S., Aron, S., Deneubourg, J.L., and Pasteels, J.M.: "Self-organized shortcuts in the Argentine ant," *Naturwissenschaften*, vol. 76, pp. 579–581, 1989.

[44] Gould, S.J.: *Evolution as fact and theory, Hen's Teeth and Horse's Toes*, W. W. Norton & Company, New York: pp. 253–262, 1994.

[45] Handl, J., Knowles, J., and Dorigo, M., "On the performance of ant-based clustering," *Frontiers in Artificial Intelligence and Applications*, vol. 104, pp. 204–213, 2003.

[46] Hardy, J., Pomeau, Y and de Pazzis, O., "Time evolution of a two-dimensional model system," *Journal of Mathematical Physics*, vol. 14, no.1746–1759, 1973.

[47] He, C., Noman, N., and Iba, H.: "An improved artificial bee colony algorithm with non-separable operator," in *Proceedings of International Conference on Convergence and Hybrid Information Technology*, 2012.

[48] Helmholtz, H.: *On the Sensations of Tone*, Dover Publications, Mineola, NY, 1954.

[49] Heiner, M., Gilbert, D. and Donaldson, R.: "Petri nets for systems and synthetic biology," *Proceedings of the 8th International Conference on Dormal Methods for Computational Systems Biology*, Lecture Notes in Computer Science, Volume

5016/2008, pp. 215–264, Springer-Verlag Berlin, 2008.

[50] Heppner, F. and Grenader, U.: *A Stochastic Nonlinear Model for Coordinated Bird Flocks*, AAAS Washington,DC, 1990.

[51] Higashi, N. and Iba, H.: "Particle swarm optimization with Gaussian mutation," *Proceedings of IEEE Swarm Intelligence Symposium (SIS03)*, pp. 72–79, 2003.

[52] Holland, J.H.: *Adaptation in Natural and Artificial Systems.* University of Michigan Press, Ann Arbor, 1975.

[53] Huberman, B.A. and Glance, N.S.: "Evolutionary games and computer simulations," *Proceedings of the National Academy of Sciences*, vol. 90, no. 16, pp. 7716–7718, 1993.

[54] Iba, H., Akiba, S., Higuchi, T. and Sato, T.: "Bugs: A bug-based search strategy using genetic algorithms," in *Proceedings of Parallel Problem Solving from Nature*, pp. 165–174, 1992.

[55] Iba, H., Higuchi, T., de Garis, H. and Sato, T.: "Evolutionary learning strategy using bug-based search," *Proceedings of the 13th International Joint Conference on Artifical Intelligence*, pp. 960–966, 1993.

[56] Iba, H., deGaris, H., and Higuchi, T.: "Evolutionary learning of predatory behaviors based on structured classifiers," *Proceedings of the Second International Conference on From Animals to Animats 2: Simulation of Adaptive Behavior*, pp. 356–363, 1993.

[57] Iba, H., Paul, T.K., and Hasegawa, Y.: *Applied Genetic Programming and Machine Learning*, Taylor & Francis, Inc., Boca Raton, 2010.

[58] Iba, H. and Noman, N.: *New Frontiers in Evolutionary Algorithms: Theory and Applications*, World Scientific Publishing Company, London, UK, 2011.

[59] Iba, H. and Aranha, C.: *Practical Applications of Evolutionary Computation to Financial Engineering: Robust Techniques for Forecasting, Trading and Hedging*, Springer-Verlag, New York, 2012.

[60] Ishiwata, H., Noman, N., and Iba, H.: "Emergence of cooperation in a bio-inspired multi-agent system," *Proceedings of Australasian Conference on Artificial Intelligence 2010(AI2010)*, LNAI, vol. 6464, pp. 364–374, Springer, 2010.

[61] Iwasa, Y., Higashi, M., and Yamamura, N.: "Prey distribution as a factor determining the choice of optimal foraging strategy," *American Naturalist*, vol. 117, pp. 710–723, 1981.

[62] Izumi, K.: "An artificial market analysis of development of market complexity," *Agent-Based Approaches in Economic and Social Complex Systems*, IOS Press, Amsterdam, pp. 47–58, 2001.

[63] Kameoka, A. and Kuriyagawa, M.: "Consonance theory, part I: Consonance of dyads," *Journal of the Acoustical Society of America*, vol. 45, no.6, pp. 1451–1459, 1969.

[64] Kameoka, A. and Kuriyagawa, M.: "Consonance theory, part II: Consonance of complex tones and its computation method," *Journal of the Acoustical Society of America*, vol. 45, no.6, pp. 1460–1469, 1969.

[65] Karaboga, D. and Basturk, B.: "A powerful and efficient algorithm for numerical function optimization: Artificial bee colony (ABC) algorithm," *Journal of Global Optimization*, vol. 39, pp. 459–471, 2007.

[66] Karaboga, D., Gorkemli, B., Ozturk. C, and Karaboga, N.: "A comprehensive survey: Artificial bee colony (ABC) algorithm and applications," *Artificial Intelligence Review*, Doi:10.1007/s10462-012-9328-0, 2012.

[67] Kauffman, S.A.: *The Origins of Order: Self-Organization and Selection in Evolution*, Oxford University Press, Oxford, UK, 1993.

[68] Kawato, M.: "From 'Understanding the brain by creating the brain' toward manipulative neuroscience," Yanagida, T., Okano, H. and Iriki, A., (Eds.), *Philosophical Transactions B*, vol. 363, pp. 2201-2214, 2008.

[69] Kirkpatrick, M.: "Sexual selection and the evolution of female choice," *Evolution*, vol. 36, pp. 1–12, 1982.

[70] Kendall, G., Yao, X. and Chong, S.-Y.: *The Iterated Prisoners' Dilemma: 20 Years On*, World Scientific Publishing Co., Inc., Singapore, 2007.

[71] Kennedy, J. and Eberhart, R.C.: "Particle swarm optimization," *Proceedings of the IEEE International Conference on Neural Networks*, pp. 1942–1948, 1995.

[72] Kennedy, J. and Eberhart, R.C.: *Swarm Intelligence*, Morgan Kaufmann Publishers, San Francisco, 2001.

[73] Kennedy, J. and Spears, W.M.: "Matching algorithms to problems: An experimental test of the particle swarm and some genetic algorithms on the multimodal problem generator," *Proceedings of the IEEE Congress on Evolutionary Computation (CEC)*, pp. 78–83, 1998.

[74] Knuth, D.E.: "Computer programming as an art," ACM Turing Award Lectures, *Communications of the ACM*, vol. 17, no.12, pp. 667–673, 1974.

[75] Kondo, S. and Asai, R.: "A viable reaction-diffusion wave on the skin of Pomacanthus, a marine angelfish," *Nature*, 376, pp. 765–768, 1995.

[76] Koza, J.R.: "Hierarchical genetic algorithms operating on populations of computer programs," *Proceedings of the Eleventh International Joint Conference on Artificial Intelligence IJCAI-89*, vol. 1, pp. 768–774, Morgan Kaufmann, San Francisco, 1989.

[77] Kusch, I. and Markus, M.: "Mollusc shell pigmentation: Cellular automaton simulations and evidence for undecidability," *Journal of Theoretical Biology*, vol. 178, pp. 333–340, 1996.

[78] Langton, C.G.(ed.): *Artificial Life*, Addison-Wesley, Boston, 1989.

[79] Levy, S.: *Hackers: Heroes of the Computer Revolution*, Anchor Press/Doubleday, Garden City, NY, 1984.

[80] Lioni, A., Theraulaz, G. and Deneubourg, J.L.: "The dynamics of chain formation in *Oecophylla longinoda*," *Journal of Insect Behavior*, vol. 14, no.5, pp. 679–696, 2001.

[81] Lioni, A. and Deneubourg, J.: "Collective decision through self-assembling," *Naturwissenschaften*, vol. 91, no.5, pp. 237–241, 2004.

[82] Luna, F. and Stefansson, B.: *Economic Simulations in Swarm: Agent-Based Modelling and Object Oriented Programming*, Kluwer Academic Publishers, Norwell, MA, 2000.

[83] Martinez, G.J.: "Introduction to Rule 110," Rule 110 Winter WorkShop, 2004 http://www.rule110.org/amhso/results/rule110-intro/introRule110.html.

[84] Metivier, M., Lattaud, C., and Heudin, J.C.: "A stress-based speciation model in lifedrop," *Artificial life VIII: Proceedings of the Eighth International Conference on Artificial Life*, pp. 121–126, 2003.

[85] Miller, G.: *The Mating Mind: How Sexual Choice Shaped the Evolution of Human Nature*, Anchor, New York, 2001.

[86] Miranda, V. and Fonseca, N.: "EPSO—Best of two worlds meta-heuristic applied to power system problems," *Proceedings of the 2002 World Congress on Computational Intelligence (WCCI2002)*, pp. 1080–1085, 2002.

[87] Mitchell, M.: "Life and evolution in computers," *History and Philosophy of the Life Sciences*, vol. 23, pp. 361–383, 2001.

[88] Mitchell, M.: *Complexity: A Guided Tour*, Oxford University Press, New York, 2009.

[89] Nagel, K. and Shreckenberg, M.: "A cellular automaton model for freeway traffic," *Journal de Physique I*, vol. 2, no.12, pp. 2221–2229, 1992.

[90] Neary, T. and Woods, D.: "P-completeness of cellular automaton Rule 110," *Proceedings of ICALP 2006 - International Colloquium on Automata Languages and Programming, Lecture Notes in Computer Science*, vol. 4051, pp. 132-143, 2006.

[91] Ninagawa, S.: "1/f noise in elementary cellular automaton rule 110," *Proceedings of the 5th International Conference on Unconventional Computation, UC06, Lecture Notes in Computer Science*, vol. 4135/2006, pp. 207–216, 2006.

[92] Noman, N. and Iba, H.: "Evolutionary computation for synthetic biology," *New Generation Computing*, vol. 31, 2013.

[93] Nowak, M.A.: *Evolutionary Dynamics: Exploring the Equations of Life*, Belknap Press of Harvard University Press, Cambridge, MA, 2006.

[94] Ohsaki, N.: "Preferential predation of female butterflies and the evolution of Batesian mimicry," *Nature*, 378, pp. 173–175, 1995.

[95] Park, J.-B., Lee, K.-S., Shin, J.-R. and Lee, K.Y.: "A particle swarm optimization for economic dispatch with nonsmooth cost functions," *IEEE Transactions on Power Systems*, vol. 20, no.1, pp. 34–42, 2005.

[96] Pollack, J.P. and Lipson, H.: "The GOLEM Project: Evolving hardware bodies and brains," *Proceedings of The Second NASA/DoD Workshop on Evolvable Hardware (EH'00)*, p.37, 2000.

[97] Poundstone, W.: *Gaming the Vote: Why Elections Aren't Fair (and What We Can Do About It)*, Hill & Wang, New York, 2008.

[98] Powell, S. and Franks, N.R.: "How a few help all: living pothole plugs speed prey delivery in the army ant *Eciton burchellii*," *Animal Behaviour*, vol. 73, no.6, pp. 1067–1076, 2007.

[99] Rajewsky, N., Santen, L., Schadschneider, A., and Schreckenberg, M.: "The asymmetric exclusion process: Comparison of update procedures," *Journal of Statistical Physics*, vol. 92, pp. 151–194, 1998.

[100] Rechenberg, I.: "Human decision making and manual control," in Willumeit, H.P., (ed.), *Evolution Strategy and Human Decision Making*, pp. 349–359, North-Holland, Amsterdam, Netherlands, 1986.

[101] Reynolds, C.W.: "Flocks, herds and schools: a distributed behavioral model," *Computer Graphics*, vol. 21, no.4, pp. 25–34, 1987.

[102] Rucker, R.: *Artificial Life Lab*, Waite Group Press, Bolinas, CA, 1993.

[103] Sandel, M.: *Justice: What's the Right Thing to Do?* Penguin, London, 2010.

[104] Sayama, H.: "Swarm chemistry," *Artificial Life*, vol. 15, no.1, pp. 105–114, 2009.

[105] Schelling, T.C.: "Dynamic models of segregation," *Journal of Mathematical Sociology*, vol. 1., pp. 143–186, 1971.

[106] Sims, K.: "Artificial evolution for computer graphics," *ACM Computer Graphics*, vol. 25, no.4, pp. 319–328, 1991.

[107] Sims, K.: "Evolving virtual creatures," *Proceedings of Computer Graphics (SIGGRAPH'94)*, pp. 15–22, 1994.

[108] Sims, K.: "Evolving 3D morphology and behavior by competition," *Proceedings of Artificial Life IV*, Brooks,R. & Maes,P. (eds.), pp. 28–39, MIT Press, Cambridge, MA, 1994.

[109] Sims, K.: Galápagos, http://www.genarts.com/galapagos/
http://www.ntticc.or.jp/, 1997.

[110] Suganthan, P.N., Hansen, N., Liang, J.J., Deb, K., Chen, Y.P., Auger, A., and Tiwari, S.: "Problem definitions and evaluation criteria for the CEC 2005 Special Session on Real-Parameter Optimization," Technical Report, Nanyang Technological University, Singapore, May, 2005.

[111] Tabor, J.J., Salis, H.M., Simpson, Z.B., Chevalier, A.A., Levskaya, A., Marcotte, E.M., Voigt, C.A., and Ellington, AD.: "A synthetic genetic edge. detection program," *Cell*, vol. 137, no.7, pp. 1272–1281, 2009.

[112] Takeuchi, Y., Iwasa, Y., and Sato, K.: *Mathematics for Ecology And Environmental Sciences (Biological and Medical Physics, Biomedical Engineering)*, Springer-Verlag, Berlin, Germany, 2007.

[113] Takashima, A., Minegishi, R., Kurabayashi, D., and Kanzaki, R.: "Construction of a brain-machine hybrid system to analyze adaptive behavior of silkworm moth," *Proceedings of 2010 IEEE/RSJ International Conference on Intelligent Robots and Systems (IROS2010)*, pp. 2389–2394, 2010.

[114] Teller, A. and Veloso, M.: "PADO: Learning tree structured algorithms for orchestration into an object recognition system," Technical Report CMU-CS-95-101, Pittsburgh, PA, USA, 1995.

[115] Takagi, H. and Ohsaki, M.: "IEC-based hearing aids fitting," *Proceedings of the IEEE International Conference on Systems, Man, and Cybernetics (IEEE SMC99)*, pp. 12–15, 1999.

[116] Takagi, H. and Iba, H.: "Interactive evolutionary computation," (special issue) *New Generation Computing*, vol. 23, no.1, pp. 113–114, 2005.

[117] Tohge, T. and Iba, H.: "Evolutionary morphology for polycube robots," *International Journal of Advanced Robotic Systems*, Frontiers in Evolutionary Robotics, pp. 567–586, 2008.

[118] Tohge, T. and Iba, H.: "Evolutionary morphology for cubic modular robot," *Proceedings of 2006 IEEE World Congress on Computational Intelligence (CEC2006)*, pp. 1995–2001, 2008.

[119] Torrance, S.B.: *The Mind and the Machine: Philosophical Aspects of Artificial Intelligence*, Ellis Horwood, Harlow, UK, 1984.

[120] Tokui, N. and Iba, H.: "Music composition with interactive evolutionary computation," *Proceedings of the 3rd Annual International Conference on Generative Art*, Milan, Italy, 2000.

[121] Unemi, T.: "SBART2.4: Breeding 2D CG images and movies, and creating a type of collage," *Proceedings of the Third International Conference on*

Knowledge-Based Intelligent Information Engineering Systems, pp. 288–291. 1999.

[122] Wakaki, H. and Iba, H.: "Motion design of a 3d-cg avatar using interactive evolutionary computation," *Proceedings of 2002 IEEE International Conference on Systems, Man and Cybernetics (SMC02)*, IEEE Press, 2002.

[123] Wallace, A.R.: *Darwinism: An Exposition of the Theory of Natural Selection, with Some of its Applications*, Macmillan, New York, 1889.

[124] Werner, G.M.: "Why the peacock's tail is so short, limits to sexual selection," in *Artificial Life V*, Langton, C.G. and Shimohara, K.(eds.) pp. 85–91, MIT Press, Cambridge, MA, 1996.

[125] Werner, G.M. and Todd, P.M.: "Too many love songs: Sexual selection and the evolution of communication," *Proceedings of the Fourth European Conference on Artificial Life (ECAL97)*, pp. 434–443, MIT Press, Cambridge, MA,1997.

[126] Wilson, E.O.: *Sociobiology: The new synthesis*, Belknap Press, Cambridge, MA, 1975.

[127] Wilkinson, G. S. "Reciprocal food sharing in the vampire bat," *Nature* 308: 181–184.

[128] Wolfram, S.: *A New Kind of Science*, Wolfram Media, Champaign, IL, 2002.

[129] Yamaguchi, M., Yoshimoto, E., and Kondo, S.: "Pattern regulation in the stripe of zebrafish suggests an underlying dynamic and autonomous mechanism," *Proceedings of the National Academy of Sciences*, vol. 104, no. 12, pp. 4790–4793, 2007.

[130] Zahavi, A. and Zahavi, A.: *The Handicap Principle: A Missing Piece of Darwin's Puzzle*, Oxford University Press, Oxford, UK, 1999.

Index

Swarm Index

Name Index

Arrow, Kenneth, 9
Arthur, Brian, 12
Axelrod, Robert, 9, 91, 95

Brooks, Rodney, 7
Burks, Arthur, 181

Conway, John, 181

Darwin, Charles, 75
Dawkins, Richard, 39

Fabre, Jean-Henri, 113
Fisher, Ronald, 76

Gell-Mann, Murray, 9
Gould, Stephen, 15

Hamilton, William, 79
Holland, John, 9, 22, 156

Kauffman, Stuart, 106
Knuth, Donald, 198

Langton, Christopher, 9, 105, 194

Rapoport, Anatol, 91
Reynolds, Craig, 141

Schelling, Thomas, 202
Searle, John, 5
Sims, Karl, 40, 106, 257

Turing, Alan, 4, 211, 253

von Neumann, John, 181

Wallace, Alfred Russel, 76

Wilson, Edward Osborne, 141
Wolfram, Stephen, 186, 193

Zahavi, Amotz, 78
Zhuangzi, 47